Lecture Notes in Computer Science 5742

Commenced Publication in 1973
Founding and Former Series Editors:
Gerhard Goos, Juris Hartmanis, and Jan van Leeuwen

Andreas Kolb   Reinhard Koch (Eds.)

Dynamic
3D Imaging

DAGM 2009 Workshop, Dyn3D 2009
Jena, Germany, September 9, 2009
Proceedings

 Springer

Volume Editors

Andreas Kolb
Computer Graphics Group
Institute for Vision and Graphics
University of Siegen
57068 Siegen, Germany
E-mail: andreas.kolb@uni-siegen.de

Reinhard Koch
Computer Science Department
Kiel University
24098 Kiel, Germany
E-mail: rk@mip.informatik.uni-kiel.de

Library of Congress Control Number: 2009933196

CR Subject Classification (1998): I.4, I.5.4, I.2.10, I.4.8, I.5, I.3.7

LNCS Sublibrary: SL 6 – Image Processing, Computer Vision, Pattern Recognition, and Graphics

ISSN   0302-9743
ISBN-10   3-642-03777-1 Springer Berlin Heidelberg New York
ISBN-13   978-3-642-03777-1 Springer Berlin Heidelberg New York

springer.com

© Springer-Verlag Berlin Heidelberg 2009
Printed in Germany

Typesetting: Camera-ready by author, data conversion by Scientific Publishing Services, Chennai, India
Printed on acid-free paper   SPIN: 12743070   06/3180   5 4 3 2 1 0

Preface

3D imaging sensors have been investigated for several decades. Recently, improvements on classical approaches such as stereo vision and structured light on the one hand, and novel time-of-flight (ToF) techniques on the other hand have emerged, leading to 3D vision systems with radically improved characteristics. Presently, these techniques make full-range 3D data available at interactive frame rates, and thus open the path toward a much broader application of 3D vision systems.

The workshop on Dynamic 3D Vision (Dyn3D) was held in conjunction with the annual conference of the German Association of Pattern Recognition (DAGM) in Jena on September 9, 2009. Previous workshops in this series have focused on the same topic, i.e., the Dynamic 3D Vision workshop in conjunction with the DAGM conference in 2007 and the CVPR workshop Time of Flight Camera-Based Computer Vision (TOF-CV) in 2008. The goal of this year's workshop, as for the prior events, was to constitute a platform for researchers working in the field of real-time range imaging, where all aspects, from sensor evaluation to application scenarios, are addressed.

After a very competitive and high-quality reviewing process, 13 papers were accepted for publication in this LNCS issue. The research area on dynamic 3D vision proved to be extremely lively. Again, as for prior workshops on this field, numerous new insights and novel approaches on time-of-flight sensors, on real-time mono- and multidimensional data processing and on various applications are presented in these workshop proceedings.

We would like to thank all the people who contributed to this event and to the workshop proceedings at hand. Special thanks go to the organizers of the DAGM, to the sponsors, to the supporting organizations, and, last but not least, to the members of the Program Committee.

September 2009

Andreas Kolb
Reinhard Koch

Organization

Executive Committee

Reinhard Koch
  (Conference Chair)   Institute of Computer Science,
                          Christian-Albrechts-University Kiel, Germany

Andreas Kolb
  (Program Chair)   Institute for Vision and Graphics,
                          University of Siegen, Germany

Erhardt Barth   Institute for Neuro- and Bioinformatics,
                          University of Lübeck, Germany

Rasmus Larsen   Department of Informatics and Mathematical
                          Models, Technical University of Denmark,
                          Copenhagen, Denmark

Program Committee

James Davis	University of California, Santa Cruz, USA
Joachim Denzler	University of Jena, Germany
Bjarne Ersboell	University of Copenhagen, Denmark
Jan-Michael Frahm	University of North Carolina, USA
Günther Greiner	University of Erlangen, Germany
Joachim Hornegger	University of Erlangen, Germany
Bernd Jähne	University of Heidelberg, Germany
Jörg Krüger	Technical University of Berlin, Germany
Klaus-Dieter Kuhnert	University of Siegen, Germany
Marcus Magnor	Technical University of Braunschweig, Germany
Roberto Manduchi	University of California, Santa Cruz, USA
Thomas Moeslund	University of Aalborg, Denmark
Thierry Oggier	Mesa Imaging, Switzerland
Abbas Rafii	Canesta, USA
Torsten Ringbeck	PMD Technologies, Germany
Sudipta Sinha	University of North Carolina, USA
Wolfgang Straßer	University of Tübingen, Germany
Christian Theobalt	University of Stanford, USA
Carlo Tomasi	Duke University, USA
Peter Xu	Massey University, New Zealand
Giora Yahav	3DV Systems, Israel

Sponsoring Institutions

PMDTechnologies GmbH, Siegen, Germany
Mesa Imaging, Zurich, Switzerland

Supporting Organizations

Eurographics Association (`www.eg.org`)
Center for Sensor Systems (ZESS) (`www.zess.uni-siegen.de`)
Deutsche Arbeitsgemeinschaft Mustererkennung (DAGM e.V.) (`www.dagm.de`)

Table of Contents

Fundamentals of ToF-Sensors

Algorithms and Data Fusion

Applications of Dynamic 3D Scene Analysis

A Physical Model of Time-of-Flight 3D Imaging Systems, Including Suppression of Ambient Light

Mirko Schmidt and Bernd Jähne

Heidelberg Collaboratory for Image Processing (HCI),
Interdisciplinary Center for Scientific Computing (IWR),
University of Heidelberg,
69115 Heidelberg, Germany
{mirko.schmidt,bernd.jaehne}@iwr.uni-heidelberg.de

Abstract. We have developed a physical model of continuous-wave Time-of-Flight cameras, which focuses on a realistic reproduction of the sensor data. The derived simulation gives the ability to simulate data acquired by a ToF system with low computational effort. The model is able to use an arbitrary optical excitation and to simulate the sampling of a target response by a two-tap sensor, which can use any switching function. Nonlinear photo response and pixel saturation, as well as spatial variations from pixel to pixel like *photo response non-uniformity* (PRNU) and *dark signal non-uniformity* (DSNU) can be modeled. Also the influence of interfering background light and on-sensor suppression of ambient light can be simulated.

The model was verified by analyzing two scenarios: The cameras response to an increasing, homogeneous irradiation as well as the systematic phase deviation caused by higher harmonics of the optical excitation. In both scenarios the model gave a precise reproduction of the observed data.

1 Introduction

Time-of-flight (ToF) 3-D cameras measure the distance of the object by determining the time τ_d which the light needs to cover the distance from a light source to the object and from the object to the sensor. With c being the speed of light, the distance d of the imaged point can be computed as

$$d = \frac{1}{2} \cdot c \cdot \tau_d \,, \tag{1}$$

where the light source is assumed to be located near the camera. ToF cameras measure this distance in each pixel - enabling the simultaneous generation of dense depth maps. Pulse-based and a phase-based ToF cameras have been realized.

The pulse based method employs discrete pulses of light which are emitted by a light source and backscattered by the object. In [3] a CMOS sensor was

R. Koch and A. Kolb (Eds.): Dyn3D 2009, LNCS 5742, pp. 1–15, 2009.

presented, which can be shuttered electronically extremely fast and with a very high precision. Another method was used in [16]: Here a conventional 2D imaging sensor is combined with a physical shutter, which can be modulated in transmissivity. This physical shutter ensures that light reaches the sensor only in a certain time window, which enables the estimation of the objects distance from the detected intensity values.

With a continuous-wave, amplitude-modulated light source the depth is determined by measuring the phase-shift between the emitted and the received optical signal. For a periodical modulation of frequency ν, the phase shift φ corresponds to a temporal shift

$$\tau_d = \frac{\varphi}{2\pi\nu}\,,\tag{2}$$

which gives the distance by using (1):

$$d = \frac{\varphi}{4\pi\nu} \cdot c\,.\tag{3}$$

To measure the phase shift between the reflected optical signal and the electronic reference signal, special sensors called *Photonic Mixing Device* (PMD) were developed. They use pixel with two quantum wells to perform a correlation of both signals. In 1995 the first sensors using such a pixelwise on-chip correlation were presented in [14] and [15]. Currently the ToF camera systems of manufacturers like *PMDTec* [12], *Mesa Imaging* [8] and *Canesta* [4] are using this approach.

The correlation function of a sinusoidal electrooptical signal $S(t)$ with an electronical reference signal $R(t)$ delayed by a phase angle Θ, assuming an angular frequency ω and a correlation range of m oscillating periods, is given by

$$S(t) = G_0 + A\sin(\omega t - \varphi)\,,$$

$$R(t) = H(\sin(\omega t + \Theta))\,,$$

$$I(\Theta) = mT\left(\frac{A}{\pi}\cos(\varphi + \Theta) + \frac{G_0}{2}\right)\,.$$

H is the Heaviside step function, meaning that $R(t)$ is assumed to be rectangular. The rectangular shaped reference signal is a good approximation of a real PMD sensor reference signal, which results from a discrete switching of the electrical field inside the sensor. G_0 is a constant describing the offset of the light source, and A is its amplitude. The full derivation is available for instance in [13].

The PMD-Sensor is able to sample the correlation function at different phasings by electronically delaying the reference signal by angle Θ. Besides the objects distance, also the intensity of the light source and of a possible background illumination are unknown. Therefore at least three measurements are necessary to estimate them.

From N equidistant sampling points located at the phase angles Θ_n, the offset a_0, amplitude a_1 and the phase shift φ of the electrooptical input signal may be estimated. Most available ToF systems use $N = 4$ samples, but also systems using more or fewer samples are feasible. As shown e.g. by [10], the optimal solution in a least square sense is given by (4) - (6) and their variance may be estimated as shown in (7) - (9). This is derived by using Gaussian error propagation and assuming an equal variance σ^2 of all acquired raw intensity values I_n.

However, in practice this simplified assumption does not hold: It does not account for a variety of factors like

- non sinusoidal light modulation $S(t)$
- non rectangle switching function $R(t)$
- non-linear photo-response
- influence of on-sensor suppression of ambient light.

Furthermore spatial variations from pixel to pixel like *photo response non-uniformity* (PRNU), *dark signal non-uniformity* (DSNU), and *dark current non-uniformity* (DCNU) (see [2]) must be considered.

$$a_0 = \frac{2}{N} \sum_{n=0}^{N-1} I_n \qquad (4) \qquad\qquad \sigma_{a_0}^2 = \frac{\sigma^2}{4} \qquad (7)$$

$$a_1 = \frac{2\pi}{N} \left| \sum_{n=0}^{N-1} I_n e^{-i2\pi(n/N)} \right| \qquad (5) \qquad\qquad \sigma_{a_1}^2 = \frac{\sigma^2}{2} \qquad (8)$$

$$\varphi = \arg \left(\sum_{n=0}^{N-1} I_n e^{-i2\pi(n/N)} \right) \qquad (6) \qquad\qquad \sigma_\varphi^2 = \frac{\sigma^2}{2a_1^2} \qquad (9)$$

$$\text{with } I_n = \frac{I(\Theta_n)}{mT}$$

1.1 Motivation and Related Work

To describe these effects, a detailed physical model of a ToF sensor is necessary, similar to that proposed by [2] for linear 2D-Sensors. With a thorough calibration this model could help improving current ToF systems.

Our goal is to simulate the data produced by a ToF camera as realistic as possible, so the optimization of existing ToF systems as well as the prediction of the characteristics of yet unavailable cameras gets feasible.

Prior models do not include all the effects discussed in the previous section. They rather focused on the simulation of whole 3D scenes. In [5] a MATLAB-based approach was chosen, were the resulting point cloud of a 3D scene is represented as superposition of single point responses. The influence of an area light source and inhomogeneous illumination of the scene was simulated by [9]. In [6,7] a simulation tool for real-time ToF data was presented. It uses the GPU to generate synthetic data for whole 3D scenes, which can be static or moving.

All these approaches focus on the simulation of ToF data for a given 3D scene. This includes issues of rendering, an adequate camera model, reflectance characteristics of the imaged objects and the position and size of the light sources. From the given ideal depth image the simulated samples are generated using a measured correlation function of a real PMD. A very simple noise model is employed to simulate the influence of noise on the acquired data. From the noisy samples, a depth image is computed.

In contrast this paper focuses on the effects influencing the quality of the generated depth image, and their origin. That means we concentrated on modeling the sensor and its noise sources very carefully. But we made only small efforts to simulate the imaging of the scene. In fact we use ideal depth maps and reflectivity maps as input for our model.

The structure of this paper is as follows: In section 2 the physical model of ToF cameras is presented, starting on a conceptual level and then explaining some necessary speed-up techniques. This very general model, which is describing most available phase based ToF systems is extended in section 3 to model a specific technique of suppressing background light, used by one manufacturer. We compared the model with a real ToF system in two different scenarios; the methods and results will be presented in section 4. Section 5 gives an conclusion and outlook.

2   A Physical Model

2.1   Assumptions

Because the investigation of errors occurring in ToF systems is not possible by regarding the isolated sensor, we have to model a whole ToF acquisition system, including a light source, the target response, the image acquisition and analysis. Our focus lies on the sensor and its noise sources, so we neglected questions about the appropriate camera model, the shape and position of the light source and scene-induced interferences like multi-reflections of the active illumination.

Our model does not simulate an area light source but employs a point light source instead (which was shown by [7] to be a good approximation). We assume the light source being located at exactly the same position as the sensor. The model uses maps containing information about the theoretical scene depth, its reflectivity and the distribution of interfering background light to cover the scene dependent quantities in a simple way.

2.2   Structure of the Model

A phase based ToF measuring setup is a system consisting of a modulated light source, a target which has some effect on the light and a ToF camera which generates data from the detected optical signal.

The model is separated into modules to ensure a high flexibility. In fig. 1 the structure and the information flow between the different modules is depicted,

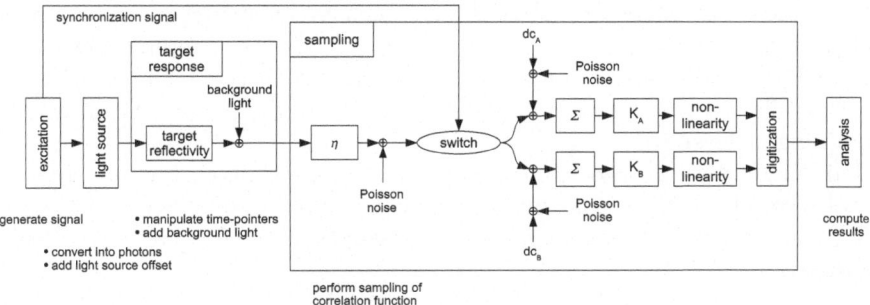

Fig. 1. Schematic representation of the model. Please see text for further information.

where a box stands for a processing unit. These units have different complexity and may consist of sub units, as it is shown for the `target response` and `sampling` module in the figure.

Excitation. The excitation module computes the function which represents the optical excitation. Furthermore a synchronization-signal is generated, which will be used in the sampling module.

Light source. Within the light source module, the excitation function is converted into a light signal. The appropriate unit of this signal is "mean number of detectable photons during one time step", so we are working with a temporal density of photons.

Target response. The target response module simulates the response of the probe. Parameters like the target's reflectivity are used here and the influence of additional (non-modulated) background light is taken into account. Because of the target's distance from the light source and ToF camera, the light signal is being shifted here against the synchronization signal.

Sampling. The sampling of the correlation function at different phasings is performed in the sampling module. Incident photons generate electrons with a certain ratio η. The conversion of "mean number of present electrons" into "present electrons" is a Poisson process, so Poisson noise is added here. A switch sorts the generated electrons into the two quantum wells A and B. Now dark current electrons are added, which are also affected by Poisson noise. The sum Σ of all collected electrons of the two taps is converted into a voltage by two distinct amplification factors K_A and K_B. This voltage is transformed by a non-linear function, which simulates the effect of pixel saturation. Both resulting voltages are digitized and these digital numbers are given out.

Analysis. From the given samples of the correlation function, the output of the camera is computed here, e.g. a phase shift and an amplitude.

All modules work on a single vector which contains the signal over time. This signal can be described as a temporal density of detectable photons, but its concrete physical meaning slightly changes between the modules.

Since we want to model phenomena which are faster than one oscillating period of the light source, we have to set the temporal sampling density to a value, which is at least one hundred times higher. So for typical integration times of 10^6 oscillating periods or more for a single depth image, we get a large vector containing at least 10^8 entries.

This might be no problem for simulating a single pixel, but our goal is to model a whole ToF sensor containing up to millions of pixels, with acceptable consumption of computing time and memory. Therefore we had to seek for optimizations.

2.3   Optimizations

In order to simulate a large number of ToF pixels simultaneously it is an interesting question which of the discussed operations are pixel-dependent and which are identical for all pixels. Because of its size, the processing of the time dependent signal vector consumes a lot of computation time and memory. Therefore it is desirable to separate it into a part which is equal for all pixels and a difference term. Since the time dependent signal is affected by noise and therefore differs randomly from pixel to pixel, this is not trivial. Fortunately it can be shown that it is possible to separate the noise in an easy way:

The process of adding Poisson noise is a function which generates random numbers which are distributed according to the Poisson distribution with a parameter λ. The Poisson distribution is given by

$$P_\lambda = \frac{\lambda^k}{k!} e^{-\lambda} .$$

The parameter λ describes the mean of the values, which is here the number of generated electrons. P_λ is the probability of detecting k electrons for a given λ.

Since the Poisson distribution is reproductive, which means that

$$X_1 \sim P_{\lambda_1}$$
$$X_2 \sim P_{\lambda_2}$$
$$\Rightarrow X_1 + X_2 \sim P_{\lambda_1+\lambda_2} , \qquad (10)$$

several time steps collecting electrons for the same tap can be grouped, and the addition of the Poisson noise can be applied only once per group. This "grouping" is exactly what the sorting module does - so it is possible to perform the switching first, and add the Poisson noise afterwards.

Each tap experiences two Poisson processes, the electrons generated from incident photons plus a certain number of dark current electrons. Both are affected by Poisson noise and may combined in order to further speed up the simulation.

By separating the time dependent signal from its noise, we have done the critical part: All other pixel dependent operations like the multiplicative factors which describe the reflectivity of the target and the quantum efficiency of each pixel, or the additive factors like the amount of incident background light and interfering dark current electrons can simply be rearranged.

Since we now know separating the time dependent signal from its noise is allowed, it is possible to compute the switching function only once and to use these values to simulate all pixels. If the excitation function is periodical and the integration time of a subframe is an integer multiple of a oscillating period, a further speedup is achieved by computing the switching function for a single oscillating period only, and multiplying the result by the number of oscillating periods per subframe.

After rearranging the model according to that explanation, it looks as shown in fig. 2.

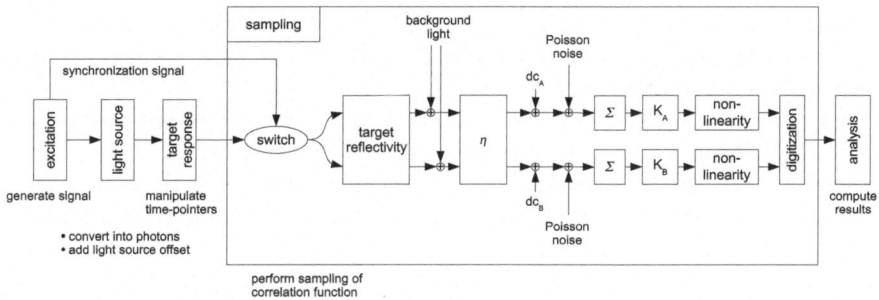

Fig. 2. Schematic representation of the model after combining Poisson processes.

By performing this sampling operation at four different phase angles $\Theta = \{0, 90, 180, 270\}^{\circ}$ of the input signal, we simulate the acquisition of four subframes, each consisting of two raw images. So we get eight raw images, as typically produced by a real ToF camera. Each pair of raw images corresponding to the same phasing, but different channels (one was taken by tap A, the other by tap B) are summed to decrease the influence of spatial inhomogeneities of the sensor. Now (6) is used to reconstruct a phase image, from which the depth image is being computed by using (3).

3   Suppression of Background Light

Due to its flexible structure, the model can easily be extended to describe even more complex systems. A very interesting question for developers and users of ToF systems is, how robust the system reacts to non-modulated background light. This interfering illumination causes an earlier saturation of the quantum wells, so less of the backscattered active light containing depth information is being detected, leading to a decreased signal-to-noise ratio (SNR).

So an interesting task for ToF manufacturers is to design systems which actively decrease the influence of non-modulated light. One system developed by *PMDTec* is called *Suppression of Background Illumination* (SBI), which is implemented in its *CamCube* ToF camera.

The manufacturer did not publish detailed information about the SBI, but it is possible to gather some information by analyzing the data produced by the camera.

3.1   Observations

When irradiating the cameras sensor with increasing intensities and analyzing the acquired intensity values of both channels A and B of a subframe of a certain angle Θ, the following behavior can be observed: For low intensities, there is a linear relation between the intensity of the light source and the sensors output. At some point, one of the channels gets saturated, i.e. there is almost no variation of the raw data while further increasing the intensity of the light source. At the same point, the output of the other raw-channel starts decreasing, while still increasing the irradiation level.

This behavior can be explained as follows: The charge stored in the two quantum wells Σ is continuously compared with a reference value $n_{SBI,start}$. As soon as the amount of stored charges of one quantum well exceeds this value, i.e. the difference n_Δ of both gets positive, a compensation process is triggered. During this process, two compensation currents are injected into the quantum wells, which contain roughly the same charge as the difference n_Δ. By doing that, the quantum well which contained more electrons at the beginning of the process is reset to $n_{SBI,start}$. The other quantum well is set to a value which is below its original value.

No important information is lost due to that process: The most interesting quantity which is reconstructible from the data is the phase shift φ, which gives the depth information. To estimate φ, only the difference of the two channels A and B is of importance, not their absolute value (see (6)).

3.2   Modeling SBI

These observations were included into the model (see fig. 3): The amount of charges of the two quantum wells Σ is continuously read into the SBI circuit. It computes the maximum of both and subtracts a reference value $n_{SBI,start}$ this difference is, if positive, multiplied by a factor C_{DK}, and an offset C_{D0} is added. These parameters were introduced to model eventual deviations from an ideal system.

The computed and transformed difference value is affected by Poisson noise. It is fed into two paths, which generate the compensation currents for the two quantum wells by multiplying with a factor C_{AK} (or C_{BK}) and adding an offset C_{A0} (or C_{B0}). The generated compensation currents are also affected by Poisson noise, which is regarded by the model.

By employing the property of the Poisson distribution of being reproductive (see (10)), the model can be optimized regarding the speed and memory consumption of a numerical implementation, which leads to the scheme shown in fig. 4. This model can be simulated much faster, because the SBI compensation currents are computed only once per quantum well, just before the read out

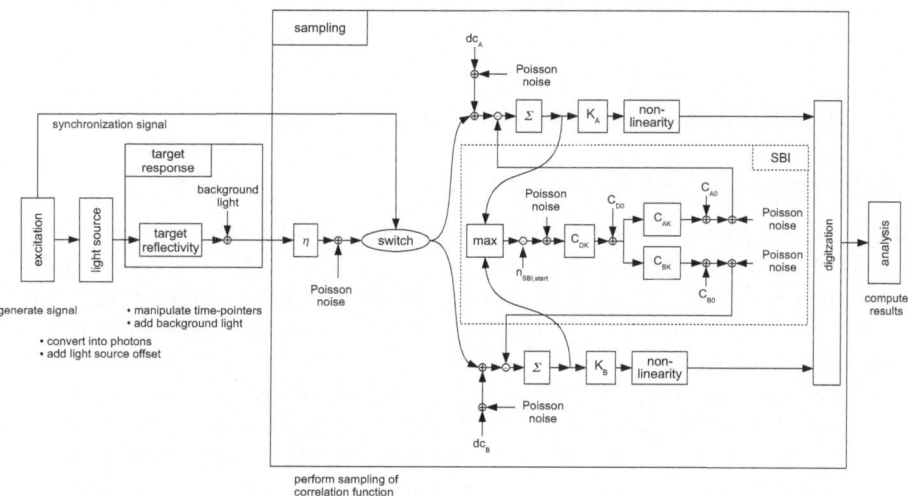

Fig. 3. Schematic representation of the model, including a SBI circuit.

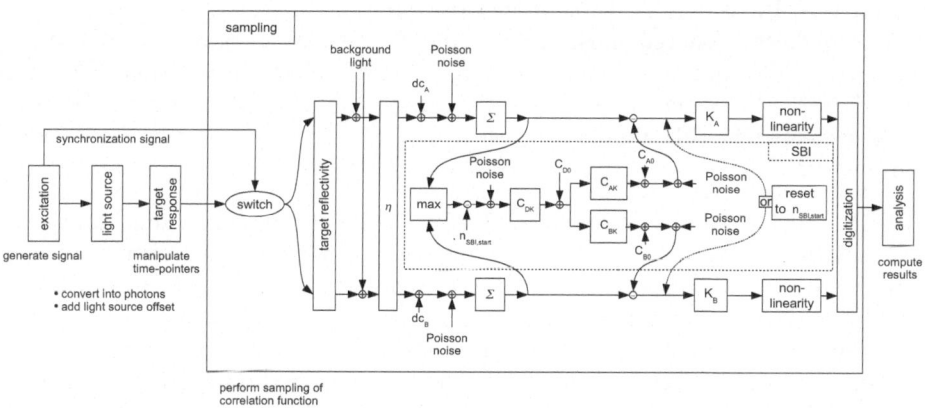

Fig. 4. Schematic representation of the model, including a SBI circuit, after combining Poisson processes.

cycle starts. In a continuous system the quantum well containing more electrons is kept on a constant level, so the additional noise caused by the SBI is canceled out by the controlling loop. We regarded that by setting the quantum well, which contained the higher number of electrons at the beginning of the SBI, to $n_{SBI,start}$ at the end of the process.

The model was implemented in *heurisko*, an image processing script language. The simulation of a ToF camera system acquiring one $1000 \times 1000\text{pixel}^2$ depth image, using four subframes, takes about $t = 10\text{s}$ on a Windows XP

Pentium 4 2, 80GHz machine. The source code was not optimized for high speed computation yet.

4   Experimental Verification of the Model

4.1   Investigation of Noise

Method. To verify the model we used a setup similar to a radiometric calibration setup for conventional 2D cameras, as described for instance in [2]: We mounted a PMD CamCube ToF camera on an integrating sphere with a calibrated light source, so that the sensor could be illuminated homogeneously with variable intensities. The radiation energy Q was varied and we observed the mean gray value μ_y and the variance σ_y^2 of the output signal of one raw image[1].

For low intensities the camera behaves like a conventional linear camera, because the SBI is not active. By applying the *photon transfer method* we were able to determine the quantities K_A, K_B and η. The idea of this technique is to exploit the fact that the detected electrons are affected by Poisson noise, which has the property of $\mu = \sigma^2$, meaning the mean of the signal is equal the its variance. So by analyzing the relation of the known number of incident photons, the generated gray values and their variance we were able to estimate the searched parameters. Please see [2,1] for further details.

The highest observed mean gray value divided by K gave the parameter $n_{SBI,Start}$. The dark currents dc_A and dc_B and their distribution were estimated from the variance of the dark signal $\sigma_{y,0}^2$. All other non-uniformities were neglected in this simulation; especially the SBI module was set to ideal parameters.

Results of noise investigation. In fig. 5 the mean gray value minus the mean dark gray value $\mu_y - \mu_{y,0}$ and variance σ_y^2 were plotted over the radiation energy Q. Also the computed corresponding quantities as a result of the simulation were plotted in the same figure. It can be seen that model gives a good explanation for the observed data. The results of the simulation and the measured quantities are very similar in the linear range up to radiation energies of $Q = 1.7 \times 10^7$ photons/pixel. At this point the SBI is activated, which causes the sharp bend in the observed and simulated data. With increasing radiation energies the model still gives a good approximation of the real ToF camera, but starts to show slight deviations. The observed variance σ_y^2 is above the simulated quantity, which was expected, because we simulated an ideal SBI module. Note that even this ideal SBI module introduces additional noise compared to a ToF system without SBI.

4.2   Investigation of Systematic Deviations

Setup. We investigated the systematic deviations of depth data predicted by the simulation and compared them to measured data. We expected a periodical

---

[1] We analyzed the channel A of subframe I_0, i.e. $\Theta = 0°$.

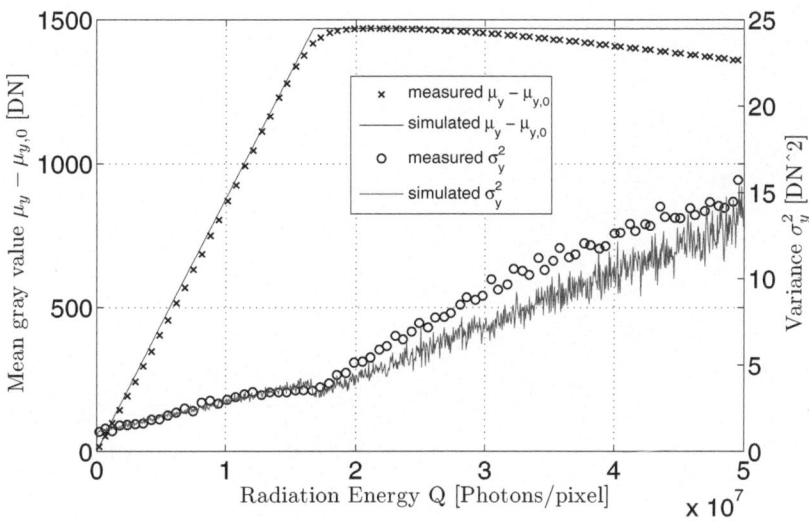

Fig. 5. Mean gray value minus mean dark gray value $\mu_y - \mu_{y,0}$ and variance σ_y^2 plotted over radiation energy Q. At $Q = 1.7 \times 10^7$ photons/pixel the SBI is activated. Please see electronic version for colors.

deviation, called "wiggling", between the measured depth and the real depth, which is caused by higher harmonics of the optical signal. A theoretical discussion of this phenomena can be found in [11].

To determine the phase deviation of the real ToF system we mounted the camera and a plane target on movable positioning tables. The light source was demounted from the camera and attached to the target, so the targets surface was irradiated from a constant distance and the backscattered light was detected by the camera. This directly illuminated target acts like a plane emitter, which has a constant irradiance independently of its distance. So the acquired depth data does not contain deviations caused by near-field effects of the optical systems (especially the light source) nor effects caused by a varying amplitude of the optical signal.

The lengthened cable from the camera to the light source introduces an additional but constant offset of the measured phase, which can easily be corrected.

We used a telephoto lens to image only a small, homogeneously irradiated area from the middle of the target. By moving the tables to certain positions we varied the distances between the active target and the camera, and analyzed the depth data of some center pixels.

To model the periodical deviations we used a fast photo diode (Femto Photoreceiver HCA-S-400M-SI-FS) and measured the shape of the optical signal. To decrease noise, we averaged the signal over 16 oscillating periods. The measured temporal modulation of the light source is plotted in fig. 6. This real shape was

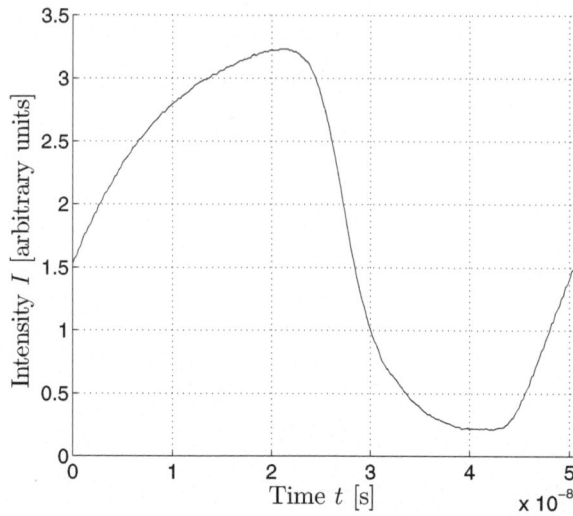

Fig. 6. Modulation of the PMD light source: intensity I plotted over time t for one oscillating period.

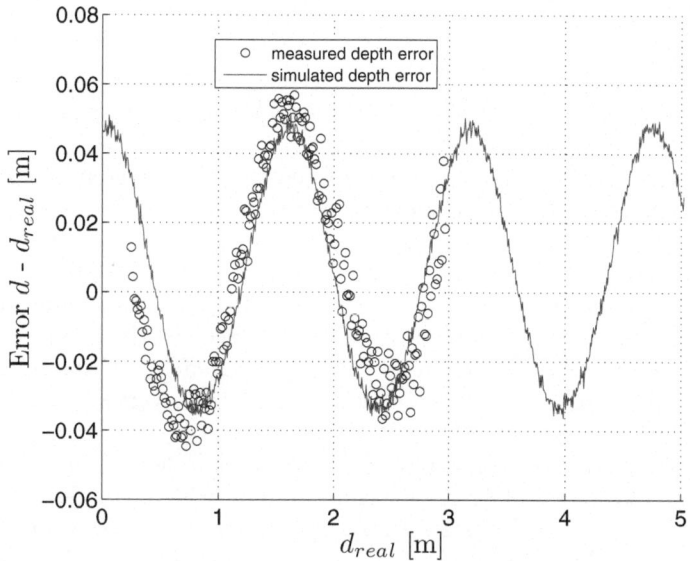

Fig. 7. Mean depth deviation of the simulated and measured distance from the real distance, plotted over the real distance.

integrated into the model and the simulation was run for a varying distance between target and camera, i.e. varying phase shifts.

Results of investigation of systematic deviations. In fig. 7 we plotted the measured and simulated depth deviations over the real depth d_{real}. The real depth can be computed from the chosen distance $d_{distance}$ between camera and target as:

$$d_{real} = 2 \cdot d_{distance} + d_{real,0}. \tag{11}$$

Note that because of the detached light source, the light has to travel the distance between target and camera only once, which explains factor 2. The distance offset $d_{real,0}$, which results from the lengthened cable and some camera internal delays of the signal, is unknown but not important for this investigation. As measured depth data we used the depth data delivered by the camera of a 10×10pixel^2 array near the optical axis.

The measured depth deviation has a periodical structure with a wavelength of a quarter of the ambiguity range, i.e. 7.5m/4. Since $d_{real,0}$ is unknown, we set it to a value which fits best to the simulated data. From fig. 7 it can be seen that the model generated a very well reproduction of the measured deviation: The wavelength and amplitude of measured and simulated deviation are in very good agreement.

5   Conclusion and Outlook

We have presented a physical model of continuous-wave ToF cameras, which offers a very high flexibility. The derived simulation gives the ability to simulate data acquired by a PMD ToF sensor with low computational effort. An arbitrary optical excitation may be used to simulate the sampling of a target response by a two tap sensor, which can use any function as switching function. All spatial parameters like the reflectivity of the target seen by a single pixel, the local amount of background light or the quantum efficiency η are treated as maps and may be specified. We have integrated an additional module to the model which simulated a circuit for suppressing ambient light.

As a verification we analyzed two scenarios: The cameras response to an increasing, homogeneous irradiation as well as the systematic phase deviation caused by higher harmonics of the optical excitation. In both scenarios the model gave a precise reproduction of the observed data.

As a next step we will model *Photo response non-uniformity* (PRNU), *dark signal non-uniformity* (DSNU) and dark current non-uniformity (DCNU) by choosing the number of dark current electrons dc, the system gain K and the nonlinearity parameters like the fullwell capacity independently for each Gate and each pixel. We are interested in modeling also pulse-based ToF systems which will require only slight modifications of the presented model.

We see our model as an important element for the development of standards to characterize and compare Time-of-Flight systems from different manufacturers. A detailed measurement and comparison of ToF systems may be found in [1].

We are also interested in using ToF cameras for *fluorescence lifetime imaging* (FLI). By modifying the presented model, e.g. changing the input module, we will be able to simulate FLI systems.

Acknowledgment

This work is partially funded by Sony EuTec Stuttgart and the BMBF project Lynkeus. We would like to thank Michael Erz for very inspiring discussions and doing great work when acquiring the data of real ToF Systems. We also would like to thank the anonymous reviewers for giving detailed constructive and thus very useful feedback. Furthermore we would like to thank Martin Schmidt for his support with the drivers of the measurement equipment.

References

1. Erz, M., Jähne, B.: Radiometric, spectrometric and range calibrations of ToF cameras. In: Koch, R., Kolb, A. (eds.) 3rd Workshop on Dynamic 3D Imaging. LNCS, vol. 5742, pp. 16–27. Springer, Heidelberg (2009)
2. EMVA Standard 1288: EMVA Standard 1288 - Standard for Characterization of Image Sensors and Cameras. European Machine Vision Association, release 3.0, draft 1e 2009 edn. (February 2009) (to appear)
3. Elkhalili, O., Schrey, O., Ulfig, W., Brockherde, W., Hosticka, B.J., Mengel, P., Listl, L.: A 64x8 pixel 3-D CMOS time-of flight image sensor for car safety applications (2006)
4. Gokturk, S.B., Yalcin, H., Bamji, C.: A time-of-flight depth sensor - System description, issues and solutions, http://www.canesta.com/assets/pdf/technicalpapers/CVPR_Submission_TOF.pdf
5. Hasouneh, F., Knedlik, S., Peters, V., Loffeld, O.: PMD based mobile node position monitoring. In: Position, Location, And Navigation Symposium, pp. 569–573. IEEE Press, New York (2006)
6. Keller, M., Orthmann, J., Kolb, A., Peters, V.: A Simulation Framework for time-of-flight Sensors. In: Proc. of the Int. IEEE Symp. on Signals, Circuits & Systems (ISSCS), vol. 1, pp. 125–128 (2007)
7. Keller, M., Kolb, A.: Real-time Simulation of time-of-flight Sensors. Simulation Practice and Theory 17, 967–978 (2009)
8. Oggier, T., Lehmann, M., Kaufmann, R., Schweizer, M., Richter, M., Metzler, P., Lang, G., Lustenberger, F., Blanc, N.: An all-solid-state optical range camera for 3D real-time imaging with sub-centimeter depth resolution (2004)
9. Peters, V., Loffeld, O., Hartmann, K., Knedlik, S.: Modeling and Bistatic Simulation of a High Resolution 3D PMD-Camera. In: EUROSIM 2007 (6th EUROSIM Congress on Modelling and Simulation), Ljubljana, Slovenia (2007)
10. Plaue, M.: Analysis of the PMD imaging system: Technical report, Interdisciplinary Center for Scientific Computing (IWR), University of Heidelberg (2006)
11. Rapp, H.: Experimental and Theoretical Investigation of Correlating TOF-Camera Systems. Diploma thesis, Interdisciplinary Center for Scientific Computing(IWR), University of Heidelberg (2007)

12. Ringbeck, T., Hagebeuker, B.: A 3D time-of-flight camera for object detection (2007)
13. Schmidt, M.: Spatiotemporal Analysis of Range Imagery. Dissertation, IWR, Fakultät für Physik und Astronomie, University of Heidelberg (2008)
14. Schwarte, R., Heinol, H.G., Xu, Z., Hartmann, K.: New active 3D vision system based on rf-modulation interferometry of incoherent light. In: Casasent, D.P. (ed.) Society of Photo-Optical Instrumentation Engineers (SPIE) Conference Series, vol. 2588, pp. 126–134 (1995)
15. Spirig, T., Seitz, P., Heitger, F.: The lock-in CCD. Two-dimensional synchronous detection of light. IEEE J.Quantum Electronics 31, 1705–1708 (1995)
16. Yahav, G., Iddan, G.J., Mandelboum, D.: 3D Imaging Camera for Gaming Application (2006),
http://www.3dvsystems.com/technology/3D%20Camera%20for%20Gaming-1.pdf

Compensation of Motion Artifacts for Time-of-Flight Cameras

Marvin Lindner and Andreas Kolb

Institute for Vision and Graphics
Computer Graphics and Multimedia Systems Group
University of Siegen, Germany
marvin.lindner@uni-siegen.de

Abstract. During the last years, Time-of-Flight sensors achieved a significant impact onto research fields in computer vision. For dynamic scenes however, most sensor's working principles lead to significant artifacts in respect to sensor or object motion – artifacts that commonly affect distance reliability and thus affect downstream processing tasks in a negative way.

We therefore introduce a compensation approach for sensors based on the Photonic Mixing Device (PMD). Out technique deals with both, lateral and axial motion artifacts. The lateral compensation tracks object motion on the level of phase images and accordingly adjusts the depth image computation in order to reduce artifacts and enhances depth reliability. The axial motion compensation is based on an axial motion estimation and a theoretical model for axial motion deviation errors. Both components utilize fast optical flow algorithms.

1 Introduction

In the context of computer vision, Time-of-Flight (TOF) sensors like the Photonic Mixer Device [1, PMD] currently became considerable alternatives to common 3D sensing devices and are already utilized for e.g. scene acquisition, obstacle detection, object tracking or gesture recognition. Most interactive applications as those mentioned before thereby deal with dynamic scenes, i.e. with moving objects or camera movements. However, due to the sensor's working principle, which is based on subsequent phase image acquisition (see sensor principle in Sec. 2.1), fast motion leads to artifacts in situations where corresponding phase images do not align properly with respect to object points and the camera's per-pixel sampling schema results in a false distance computation.

For this purpose, we present a new compensation approach for motion artifacts, that uses optical flow to break up the fixed sample schema and allows the realignment of corresponding phase images by tracking individual object points over time. We further introduce an axial motion compensation, that deals with motion along the viewing direction. In boundary regions, axial motion leads to mismatching phase values as well. Additionally, we give a theoretical analysis of the effects of axial motion to the distance measurement and introduce a correction scheme based on axial motion estimation. A detailed description of the

R. Koch and A. Kolb (Eds.): Dyn3D 2009, LNCS 5742, pp. 16–27, 2009.

compensation approach, i.e. the necessary processing steps to make optical flow applicable for phase images, is given in Sec. 3. Prior to that, a principle description about TOF sensing and motion artifacts is given in Sec. 2. Implementation details are finally discussed in Sec. 4 along with two exemplary results in Sec. 5.

2   Technical Background

2.1   PMD-Based TOF-Principle

The main component of common PMD/TOF-cameras is represented by a CMOS chip consisting of so-called smart pixels, each measuring individual distance information for an observed scene [2,3,4]. The underlying measurement process itself is based on the TOF principle as illustrated in Fig. 1.

The illuminating units of the camera emit intensity modulated near infrared light (NIR). The modulated illumination, driven by the internal signal s, is reflected at the object surfaces and the resulting optical signal r is detected by the corresponding smart pixel of the PMD-sensor. Each smart pixel finally determines the correlation c between the incident optical signal r and the internal reference signal s additionally shifted by an internal phase offset τ:

$$c(\tau) = r \otimes s = \lim_{T \to \infty} \int_{-T/2}^{T/2} r(t) \cdot s(t + \tau) \, dt. \tag{1}$$

In most approaches sinusoidal signals are assumed,

$$s(t) = \cos(\omega t), \qquad\qquad r(t) = k + a \cos(\omega t + \phi) \tag{2}$$

where $\omega = 2\pi f$ is the modulation frequency, a is the amplitude of the incident optical signal and ϕ is the phase offset relating to the object distance, finally yielding

$$c(\tau) = \frac{a}{2} \cos(\omega \tau + \phi). \tag{3}$$

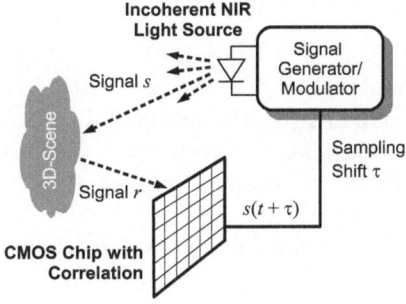

Fig. 1. The principle of PMD/TOF-measurement

By sampling the correlation function four times $I_i = c(i \cdot \pi/2), i = 1 \ldots 4$, i.e. taking four sequential phase images with an internal phase delay of $\tau_i = i \cdot \pi/2$, and using simple trigonometry, we can determine a pixel's phase shift $\phi \in [0, 2\pi]$, the correlation amplitude a and the incident light intensity h as

$$\phi = \text{atan2}\,(I_3 - I_1, I_0 - I_2) + \pi, \qquad h = \tfrac{1}{4} \sum_{i=0}^{3} I_i,$$

$$a = \tfrac{1}{2}\sqrt{(I_3 - I_1)^2 + (I_0 - I_2)^2}. \tag{4}$$

Finally, the measured distance to the according object region m is given by $m = c_{\text{light}}/4\pi f \cdot \phi$, where $c_{\text{light}} \approx 3 \cdot 10^8 \tfrac{\text{m}}{\text{s}}$ represents the speed of light and f stands for the signal's modulation frequency. Commonly a modulation frequency of 20 MHz is used, resulting in an unambiguous distance range of 7.5 m. Due to various factors, m is not the extact object distance [5]. Thus, we differentiate between the measured distance m and the true distance d.

Todays devices offer a resolution of up to 204×204 px providing distance information at 25 Hz. Due to an automatic suppression of background light, some TOF-cameras are also suitable for outdoor applications.

2.2   Motion Artifacts

Motion Artifacts typically occur where objects or the camera itself move, while the consecutive phase images I_0 - I_3 are taken. They arise from unmatching phase values during the demodulation process and are more extensive the faster the object moves or the longer the integration time is. We distinguish between three error sources:

Lateral Motion which primary result in the mixture of foreground and background phase values at the boundary of moving objects.

Axial Motion which describe motion along the viewing direction and introduce additional phase changes due to non-constant object distance.

Texture Changes which occur for objects of varying reflectivity and result in unmatching phase values even if the object distance does not change for a given pixel.

A theoretical investigation of discontinuity and texture related motion artifacts has been published by Schmidt [6]. He assumes that both even as well as odd correlation samples are taken at the same time, i.e. are related to the same reflectivity, and describes the theoretical impact of varying intensity onto the resultant distance information.

Nevertheless, no specific technique to compensate TOF camera related motion artifacts has been published yet. The only known approach by Lottner et. al [7] is based on the bilateral filtering of monocular combined 2D/3D TOF images. However, this approach equals rather an edge preserving image smoothing than a real motion compensation.

3   Motion Compensation Based on Optical Flow

As the problem basically arises from unmatching phase images due to non-constant distance and texture change, our main idea is to discard the fixed per-pixel sampling schema by tracking individual surface points in all phase images and select the correct object location in the phase image to determine the phase values for the final distance calculation. The most universal approach in this context is optical flow [8,9], as it works on a per pixel basis and allows unrestricted motion as well as deformation.

3.1   Adopting TOF-Data to Optical Flow

Applying optical flow in combination with variable sampling positions however requires the following conditions that are not fulfilled by phase images immediately.

Brightness Constraint. Tracking objects between subsequent images is based on the assumption that corresponding surface points appear with the same intensity in subsequent images.

Pixel Homogeneity. Applying the demodulation at different pixel locations requires matching raw values, i.e. a homogeneous sensor behavior, in order to get the correct phase shift.

Both conditions are not directly met by PMD-based TOF-cameras. Thus, we next will discuss solutions to this issues in the following paragraphs.

Brightness Constraint. An important precondition of optical flow is the assumption of constant intensity values between consecutive images. Unfortunately, by taking a look a the phase images I_i, it becomes obvious that objects appear differently in each phase image due to the internal phase shift and the applied convolution (see Fig. 2, top).

Fig. 2. PMD Phase Image $A - B$ (top row) and their according intensity images $A + B$ (bottom row)

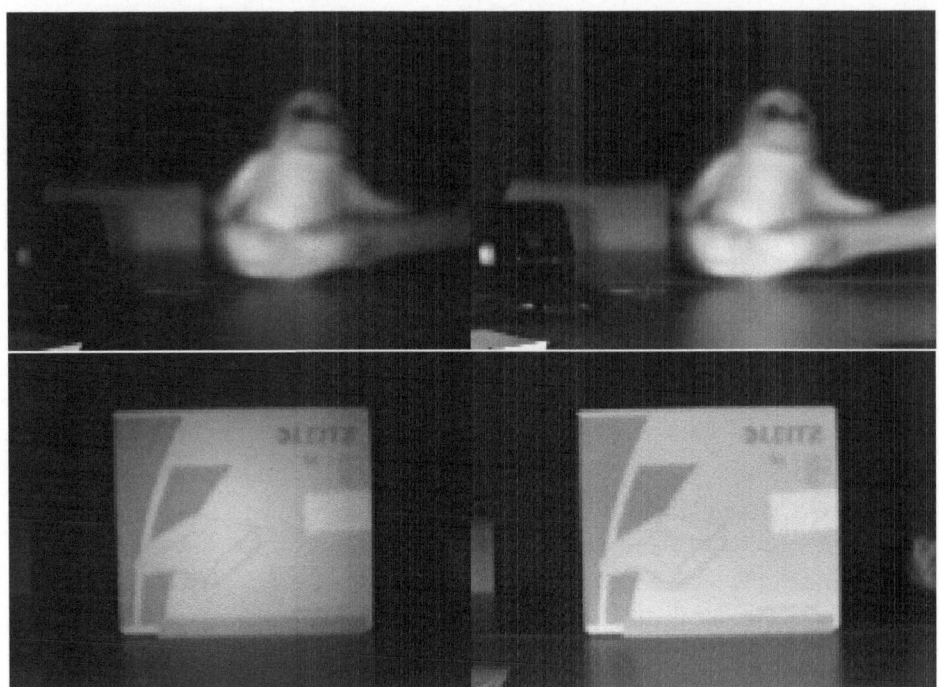

Fig. 3. PMD intensity image before (left) and after homogeneity adjustment (right). Note the strongly varying intensity of the hand/arm without correction.

However, PMD-based TOF sensors measure two raw images at a time, i.e.

– the shifted reference signal yielding $A_i = c(i \cdot \pi/2\omega)$ as well as
– the inverted signal yielding $B_i = c(i \cdot \pi/2\omega + \pi)$.

Both raw images are internally subtracted to form the actual phase image $I_i = A_i - B_i$ in order to reduce production-specific pixel behavior. Accordingly, the pixel intensity is given analog to Eq. 4 by

$$h = 1/8 \cdot \sum_{i=0}^{3}(A_i + B_i). \tag{5}$$

As both signals are inverse to each other, the absolute amount of incident light – and thus the total intensity I_i^+ – for a phase image can be computed by the sum of its raw images (see Fig. 2, bottom), making optical flow estimation finally applicable.

Pixel Homogeneity. In practice, pixel gain differences as well as a radial light attenuation towards the image border affects the phase values (see Fig. 3, left column). Concerning the fixed sampling scheme, these individual pixel characteristics are simply ignored. For the realigned demodulation, however, the inhomogeneity not only influences the optical flow estimation in a negative way by violating the constant intensity assumption, it also leads to non-matching raw values during the realigned demodulation since different pixel locations are getting combined.

Thus, in order to adjust pixel inhomogeneities and consequently improve the motion compensation, we adapted the intensity-value standardization by Sturmer [10] to work on raw images. Analog to Sturmer's intensity standardization, planar reference images of different reflectivity have been acquired and serve as basis for the determination of an appropriate correction term. In our case, the final raw value adjustment has been obtained via a pixel-wise function fit

$$f_A(A_i) = \tilde{A}_i, \ f_B(B_i) = \tilde{B}_i \ \text{with} \ i = 0\ldots3 \tag{6}$$

minimizing

$$\sum_{i=0}^{3} (\tilde{A}_i + \tilde{B}_i) = h_{\text{ref}}. \tag{7}$$

Here, the reference intensity h_{ref} is given by the brightest pixel observing a homogeneous planar surface. In presence of noise, the reference intensity should be determined using a small neighborhood.

As the correction $f_X(X_i)$ should be smooth and monotonically increasing, we decided to fit a function of logarithmic form, i.e. $f_X(X_i) = a\sqrt{X_i + b} + cX_i + d$, yielding the homogenization results as shown on the right side of Fig. 3.

3.2   Axial Motion

While optical flow handles error sources caused by lateral object shifts, the impact of axial motion still needs further investigations. Unlike lateral motion which simply results in a displacement of corresponding phase values, axial motion introduces additional phase changes due to the varying object displacement.

Assuming a uniform axial motion with an axial depth difference $\kappa = d_3 - d_0$ between phase image I_0 and I_3, the theoretical correlation samples are given analog to Eq. 3 by

$$I_i = \frac{a}{2} \cos\left(i \cdot \frac{\pi}{2} + \phi_d + \kappa_i \cdot \pi/3.75\right), \ \kappa_i = i \cdot \frac{\kappa}{3} \tag{8}$$

As a result, the demodulation schema from Eq. 4 gives the phase offset

$$\phi_m = \arctan \frac{\sin(\phi_d + a\kappa) + \sin\left(\phi_d + \frac{1}{3}a\kappa\right)}{\cos(\phi_d) + \cos\left(\phi_d + \frac{2}{3}a\kappa\right)}$$
$$= \arctan \frac{\sin\left(\phi_d + \frac{2}{3}a\kappa\right)}{\cos\left(\phi_d + \frac{1}{3}a\kappa\right)}, \tag{9}$$

where $a = \pi/3.75$. Thus for known κ and $\phi_m = a \cdot m - \pi$, the true phase offset ϕ_d is given by

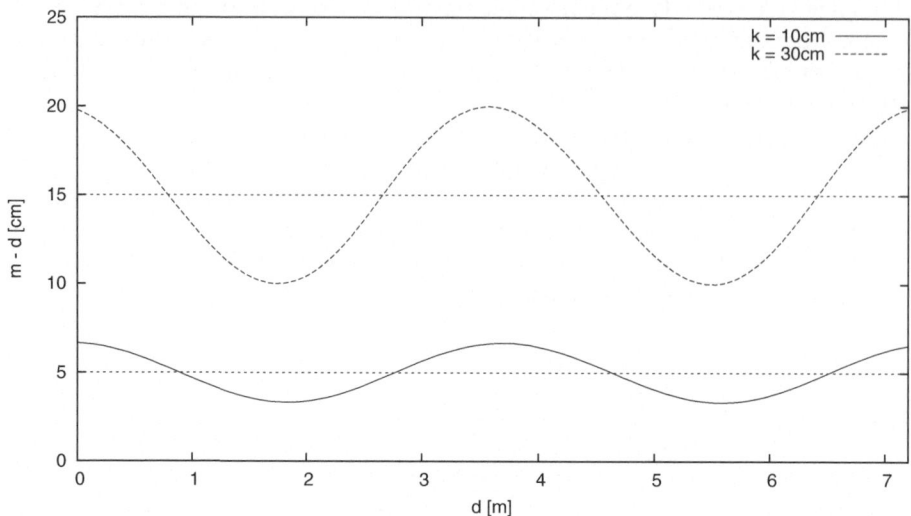

Fig. 4. Theoretical distance deviation for a total displacement of $\kappa = 10$cm and $\kappa = 30$cm

$$d = \text{atan2}\left(\sin(\phi_m) - \cos(\phi_m)\sin\left(\frac{a\kappa}{3}\right),\ \cos(\phi_m)\cos\left(\frac{a\kappa}{3}\right)\right) + \pi - \frac{a\kappa}{3}. \quad (10)$$

By taking a look on the deviation between the theoretical measured distance $d(\phi)$ and its ground truth distance m (cmp. Fig. 4), it become s obvious that the deviation smoothly fluctuates about $\kappa/2$ with an amplitude of $\kappa/6$ not causing any critical distance jumps.

Regarding a common integration time of 15 ms per raw image, the displacements in Fig. 4 already imply a rather high velocity of 8 km/h and 24 km/h respectively. For common scenarios axial motion itself therefore generally has less significant impact on the distance accuracy but is still in the range of centimeters. Additionally, a resizing of object contours occur, which are already handled by our optical flow based pixel alignment.

4   Implementation Details

The idea of motion compensation via flow estimation is finally realized using the optical flow implementation of Zach et. al [11]. Zach's GPU-based implementation allows discontinuity preserving TV-L1 flow estimation in real time and currently hold the second place of the Middlebury's optical flow ranking [12]. As the quality of the motion compensation relies on the underlying flow estimation, our choice should give the best results in respect to accuracy and runtime currently possible. A complete system overview is given in Fig. 5, consisting of both a lateral as well as an axial motion compensation.

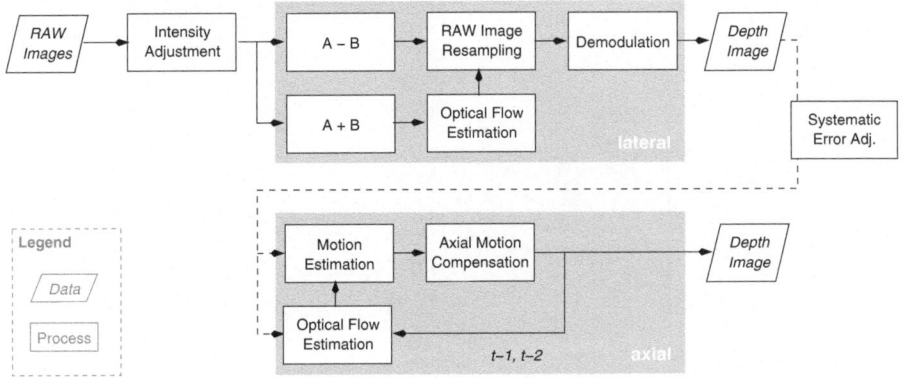

Fig. 5. System Overview. The motion compensation consists of two consecutive modules: the lateral motion and the axial motion compensation. An additional systematic error correction is necessary to adjust demodulation errors before a correct velocity estimation can be performed.

4.1   Lateral Motion Compensation

The lateral compensation is done by estimating the optical flow between the phase intensity images $I_1^+ - I_3^+$ and I_0^+ and an according resampling of the corresponding phase images. Before, all raw images are adjusted according to Sec. 3.1 to eliminate pixel inhomogeneities. The resampled phase images are then further processed by the fixed standard demodulation schema in order to calculate the final distance information (upper part of Fig. 5).

4.2   Axial Motion Impact

For depth image sequences an additional compensation approach for axial motion errors is applied. Here, object velocities are estimated via the previous two corrected depth images. Knowing the velocity of a surface point, its theoretical deviation as well as its according correction can be derived from Eq. 9. Once more, optical flow is used to track individual surface points and gives an according distance mapping for the velocity estimation between depth images.

Note that for PMD cameras, the camera's systematic distance deviation [13] must be corrected first. Otherwise, the demodulation error will lead to incorrect object displacements.

For a single depth image, e.g. when motion estimation is not possible, a synthesized intensity image halfway between I_1^+ and I_2^+ rather than I_0^+ should be considered as optical flow reference. Otherwise, the object contour will not match its distance information due to the distance shift of $\kappa/2$, causing the object to shrink or expand.

Fig. 6. *Scene 1* before (top) and after motion compensation (middle) as well as the static reference scene (bottom). Note the motion artifacts due to texture changes of the box surface.

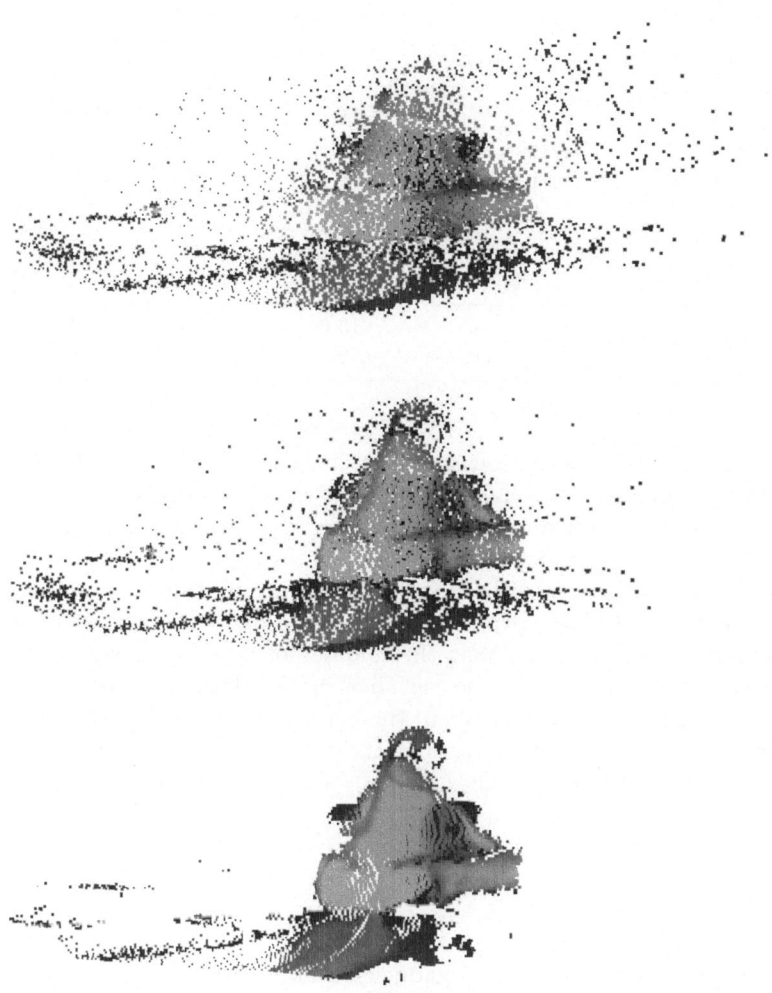

Fig. 7. *Scene 2* before (top) and after motion compensation (middle) with additional outlier removal and bilateral filtering (bottom). Note the reduction of artifacts as well as the sharpened object features.

5   Results

The motion compensation has been tested on two exemplary scenes and achieved a frame rate of 10 fps. Whereas *Scene 1* consists of a simple moving box yielding only lateral motion artifacts (see Fig. 6), *Scene 2* consist of a more complex moving soft toy, allowing additional deformation (see Fig. 2 and 7). In both cases, motion artifacts occurred due to contour or texture changes and their corresponding phase mismatch.

Table 1. Statistical analysis of *Scene 1* stating the number of object and background pixel as well as the corresponding average distance and its variance

		Static	Motion	Adjusted
	Background	7326	6904	7213
Pixel Count	Object	8521	6609	8458
	Outliers	3033	5367	3209
Mean Dist. [mm]	Background	2404	2405	2405
	Object	1271	1272	1272
Variance [mm]	Background	13.6	14.6	14.0
	Object	8.5	9.6	7.9

By applying our compensation approach, most of the arising artifacts has been satisfactorily removed. Especially the comparison with the static scene, shows that the re-sampled box for example matches the reference scene very well in distance information and object size.

A statistical evaluation of the box scene is given in Tab. 1. The detection of box and background pixels (wall) is done using a clustering approach based on plane fitting. It can be seen, that the number of detected pixels is extremely close to the static situation. The variance in the data decreases by applying the motion compensation. No shift in the average distance for the detected box or background pixels occur. Note, that the texture-related errors in the object region are not captured by the variation, since these pixels are classified as outliers.

6   Conclusions

We presented a technique for the compensation of PMD camera related motion artifacts. The compensation approach is based on the distinction between lateral and axial motion effects and treats both effects by applying optical flow results to consecutive phase images (handling lateral motion and object resizing in respect to axial motion) as well as depth images (estimation of axial velocity). The quality and performance of the compensation therefore mainly depends on the accuracy of the underlying optical flow algorithm. By using the GPU based implementation of Zach et. al [11], we have been able to produce satisfying results at a frame rate of about 10 fps.

Acknowledgment

This work is partly supported by the German Research Foundation (DFG), KO-2960/5.

References

1. PMD Technologies, http://pmdtec.com
2. Kraft, H., Frey, J., Moeller, T., Albrecht, M., Grothof, M., Schink, B., Hess, H., Buxbaum, B.: 3D-camera of high 3D-frame rate, depth-resolution and background light elimination based on improved PMD (photonic mixer device)-technologies. In: OPTO (2004)
3. Lange, R.: 3D time-of-flight Distance Measurement with Custom Solid-State Image Sensors in CMOS/CCD-Technology. PhD thesis, University of Siegen (2000)
4. Xu, Z., Schwarte, R., Heinol, H., Buxbaum, B., Ringbeck, T.: Smart pixel – photonic mixer device (PMD). In: Proc. Int. Conf. on Mechatron. & Machine Vision, pp. 259–264 (1998)
5. Kolb, A., Barth, E., Koch, R., Larsen, R.: Time-of-flight sensors in computer graphics. In: Proc. Eurographics (State-of-the-Art Report) (2009)
6. Schmidt, M.: Spatiotemporal Analysis of Imagery. PhD thesis, University of Heidelberg (2008)
7. Lottner, O., Sluiter, A., Hartmann, K., Weihs, W.: Movement artefacts in range images of time-of-flight cameras. In: International Symposium on Signals, Circuits & Systems - ISSCS 2007, vol. 2, pp. 117–120 (2007)
8. Horn, B., Schunck, B.: Determining Optical Flow. Jones and Bartlett Publishers, Inc. (1992)
9. Lucas, B.D., Kanade, T.: An iterative image registration technique with an application to stereo vision. In: Proceedings of the 7th International Joint Conference on Artificial Intelligence (IJCAI 1981), pp. 674–679 (1981)
10. Sturmer, M., Penne, J., Hornegger, J.: Standardization of intensity-values acquired by time-of-flight-cameras. In: IEEE Computer Society Conference on Computer Vision and Pattern Recognition Workshops, 2008. CVPRW 2008, pp. 1–6 (2008)
11. Zach, C., Pock, T., Bischof, H.: A duality based approach for realtime tv-l1 optical flow. In: Hamprecht, F.A., Schnörr, C., Jähne, B. (eds.) DAGM 2007. LNCS, vol. 4713, pp. 214–223. Springer, Heidelberg (2007)
12. Middlebury Database: http://vision.middlebury.edu/flow/
13. Lindner, M., Kolb, A.: Lateral and depth calibration of PMD-distance sensors. In: Bebis, G., Boyle, R., Parvin, B., Koracin, D., Remagnino, P., Nefian, A., Meenakshisundaram, G., Pascucci, V., Zara, J., Molineros, J., Theisel, H., Malzbender, T. (eds.) ISVC 2006. LNCS, vol. 4292, pp. 524–533. Springer, Heidelberg (2006)

Radiometric and Spectrometric Calibrations, and Distance Noise Measurement of ToF Cameras

Michael Erz and Bernd Jähne

Heidelberg Collaboratory for Image Processing (HCI),
Interdisciplinary Center for Scientific Computing (IWR),
University of Heidelberg, 69115 Heidelberg, Germany
{michael.erz,bernd.jaehne}@iwr.uni-heidelberg.de

Abstract. This paper proposes to extend the EMVA 1288 standard to characterize the properties and noise of image sensors for ToF cameras. The concepts for radiometric and spectrometric sensitivities were extended for intensity images recorded by lock-in pixels. The characterization of the distance information was performed by describing the phase shift analogous to intensities. Results of sensitivity and noise measurements are presented for two ToF cameras: PMDTec CamCube and MESA Imaging SR3101. Both cameras had no intrinsic filter, so the quantum efficiency could be measured from UV to IR. The noise in the phase measurement could be related to the noise in the intensity.

1   Introduction

1.1   Background

Recently, the European Machine Vision Association (EMVA) has released the standard 1288 [1] to define a unified method to measure, compute and present specification parameters and characterization data for image sensors relating to conventional cameras. The standard covers digital cameras with linear photo response characteristic. The specifications include sensitivity, noise, dark current, sensor inhomogeneities and trigger behavior. To the best of our knowledge, no standard procedure is available, to measure the performance of lock-in pixels and to relate the statistical uncertainty and systematic errors in the distance measurements to the statistical properties of the sensor. Concerning sensor theory and calibration, there is a lot of work done in [3,4,5,6,7,8]. The theoretical response of a PMD sensor for a given 3D scene and given correlation function was simulated in [9,10]. But gray values of the four phase images were here not investigated. In this paper, we propose to extend the EMVA 1288 for lock-in pixels in ToF cameras with the intention to provide a concise description of measurement process and specification information for an objective comparison and evaluation.

R. Koch and A. Kolb (Eds.): Dyn3D 2009, LNCS 5742, pp. 28–41, 2009.
© Springer-Verlag Berlin Heidelberg 2009

We investigated the applicability of the EMVA 1288 for lock-in pixels with two potential wells. Such pixels provide the phase shift in the modulated optical signal additionally to the intensity information. The phase information was examined analogous to the intensity information. The characterization of the sensitivity was transferred from the standard to the two potential wells. We had a special opportunity to use two ToF cameras without intrinsic filters and with detachable light sources, thus a full radiometric and spectrometric calibration was possible. Due to the technical features of the cameras the phase noise characterizations were performed with two different setups but equal evaluation procedures.

In terms of notation: the symbols μ_x and σ_x^2 will denote the mean value und variance of a quantity x.

1.2   Radiometry and Quantum Efficiency

The intensity characterization is based on measurements of the radiometric and spectrometric sensitivities. The radiometry deals with electromagnetic radiation at wavelengths in visible range and also in the UV and the IR range. The fundamental quantities in the radiometry are inter alia the irradiance $E[\mathrm{W/m^2}]$ and the photon irradiance $E_p[\mathrm{photons}/(\mathrm{s\,m^2})]$, which can be converted into each other as

$$E_p = \frac{\mu_p}{At_{exp}} = \frac{E}{hc/\lambda} \tag{1}$$

with the average number of photons μ_p, pixel area A, exposure time t_{exp}, the speed of light c, the Planck's constant h and wavelength of the light λ.

The characteristic quantities are the overall system gain K [DN/electrons] (digits per electrons) and the quantum efficiency η with units electrons/photons. The quantum efficiency describes the ratio of the average number of electrons μ_e produced in the optical sensitive area to the average number of incident photons:

$$\eta(\lambda) = \frac{\mu_e}{\mu_p} \ . \tag{2}$$

The product

$$R = K\eta \tag{3}$$

is the responsivity with units DN/photons and describes how the intensity depends on the irradiance. Information about the quantum efficiency should include the whole sensitive spectral range of the detector material.

The lock-in sensor consists of two potential wells A and B unlike conventional sensors with just one. Thus, the average number of electrons accumulated during the exposure time is the sum of electrons from both potential wells

$$\mu_e = \mu_e^A + \mu_e^B \ . \tag{4}$$

In principle, each potential well has also it's one gain, thus the overall system gains K^A and K^B can be different.

The main question of interest is: how does the gray value and noise behave at different irradiances E resp. E_p, and how is the quantum efficiency at different wavelengths λ?

The characterization procedure is presented in sec. 2.1. The measurement setups are described in sec. 3.1 and 3.2, the results in sec. 4.

1.3   Time-of-Flight and Phase Shift

In ToF measurements with a modulated light source the distance is calculated from the temporal delay of the signal. Therefore, the characterization of the phase shift can be representative for the distance characterization. The phase shift image φ [rad] was calculated from the four raw images recorded by the camera

$$
\begin{aligned}
y^{A'} &\text{ at } \ \ 0° \ , \\
y^{B'} &\text{ at } 180° \ , \\
y^{A''} &\text{ at } \ 90° \text{ and} \\
y^{B''} &\text{ at } 270°
\end{aligned}
$$

in the sinusoidal optical signal with

$$
\varphi = \arctan\left(\frac{(y^{B'} - y^{B'}_{\text{dark}}) - (y^{A'} - y^{A'}_{\text{dark}})}{(y^{B''} - y^{B''}_{\text{dark}}) - (y^{A''} - y^{A''}_{\text{dark}})} \right) \ . \tag{5}
$$

A more general formula for computing φ from N samples can be found in [11].

Here the question is: how does noise in phase shift images σ^2_φ [rad^2] behave at different distances and how does it relate to the noise in gray values?

The characterization procedure is presented in sec. 2.2. The two measurement setups for both cameras are described in sections 3.3 and 3.4, the results in sec. 4.

2   Methods

The characteristic quantities which should reflect the performance of the distance measurement are not the primary goal of this paper. The systematic errors, intensity related errors, integration time related errors, flying pixels or problems with motion compensation were not investigated. The objective of this paper is the characterization of the radiometric, spectrometric and noise properties of the lock-in pixel. Only the statistical errors in the measurement of the phase shift and their relation to the statistical errors in gray values were included. The systematic errors like

- wiggling error,
- a constant phase deviation per pixel,

- overexposure and exposure time dependent errors,
- phase drift,
- near field errors due to the extended illumination

were already sophisticated. Before the definition of the extended version of the EMVA 1288 standard a set of important figures and/or specific sensor characteristic quantities describing the performance of the lock-in sensor must be established.

The radiometric and spectrometric characterizations of the cameras are made according to the EMVA 1288 standard [1,2], which is based on the photon transfer method [12,13]. The next section describes this method briefly.

2.1   Radiometric and Spectrometric Sensitivities

A digital image sensor converts photons hitting the optical active pixel area in a certain time interval into a digital number. A linear signal model is assumed for this process. That means: the average digital signal μ_y [DN] is proportional to the average number of electrons μ_e being produced by photo effect in the sensor material. Since a lock-in pixel has two potential wells this process should be described for each one separately. The following equation can be used

$$\mu_y = \mu_{y,\mathrm{dark}} + \mu_e K \overset{(2)}{=} \mu_{y,\mathrm{dark}} + \mu_p \eta(\lambda) K \ , \tag{6}$$

with the dark signal $\mu_{y,\mathrm{dark}}$ present without light and the overall system gain K. Generally, the dark signal depends on the exposure time and the ambient temperature.

In [1] is shown, that the variance of gray values σ_y^2 [DN^2] can be related to the measured average digital number μ_y as

$$\sigma_y^2 = \sigma_{y,0}^2 + (\mu_y - \mu_{y,\mathrm{dark}})K \ . \tag{7}$$

The offset $\sigma_{y,0}^2$ depends on the dark signal variance and quantization noise.

For the measurements the camera was illuminated without the optics at different irradiances E_p. The number of photons per pixel was calculated with Eq. (1). The gray value averaged over all N pixels at certain irradiation levels is computed from two recorded images $y^{(1)}$ und y^2 as

$$\mu_y = \frac{1}{2N} \sum_{ij}^{N} (y_{ij}^{(1)} + y_{ij}^{(2)}) \ . \tag{8}$$

The variance of gray values is computed as a mean of the squared difference of the two images. We assume here that the noise is stationary and homogeneous.

$$\sigma_y^2 = \frac{1}{2N} \sum_{ij}^{N} (y_{ij}^{(1)} - y_{ij}^{(2)})^2 \ . \tag{9}$$

The responsivity $R = K\eta$ is the slope in Eq. (6) for μ_p as argument. The overall system gain K is the slope in Eq. (7) with $(\mu_y - \mu_{y,\text{dark}})$ as argument. The quantum efficiency is then equal to $\eta(\lambda) = R/K$. For the computation of K values of both potential wells were fitted separately.

The signal-to-noise ratio

$$\text{SNR} = \frac{\mu_y - \mu_{y,\text{dark}}}{\sigma_y} \tag{10}$$

was computed for data and for linear fits of the data (SNR_{fit}).

2.2 Phase Shift and Distance

The same principle was used for the calculation of the average phase shift μ_φ [rad] and the variance σ_φ^2 [rad^2]. The phase images were calculated with Eq. (5). Like H. Rapp et al. [6] suggested, 100 images were recorded for mean calculation. Due to nonuniformities in the gray values each pixel has a different phase offset φ_{offset}, which was subtracted from the phase images. Then the average phase shift results from

$$\mu_\varphi = \frac{1}{N} \sum_{ij}^{N} \left(\frac{1}{100} \sum_{k=1}^{100} \varphi_{ij}^{(k)} - \varphi_{ij,\text{offset}} \right) = \frac{1}{N} \sum_{ij}^{N} (\bar{\varphi}_{ij} - \varphi_{ij,\text{offset}}) \tag{11}$$

and the variance averaged over all pixels is

$$\sigma_\varphi^2 = \frac{1}{N} \sum_{ij}^{N} \left(\frac{1}{100} \sum_{k=1}^{100} \left(\varphi_{ij}^{(k)} - \bar{\varphi}_{ij} \right)^2 \right) . \tag{12}$$

The average gray value and the variance were computed as

$$\mu_y = \frac{1}{100N} \sum_{ij}^{N} \sum_{k=1}^{100} y_{ij}^{(k)} \tag{13}$$

and

$$\sigma_y^2 = \frac{1}{100N} \sum_{k=1}^{100} \sum_{ij}^{N} (y_{ij}^{(k)} - \mu_y)^2 . \tag{14}$$

The variance of the phase shift can also be calculated from the variance of the gray values and the modulation amplitude m_0[DN] as

$$\sigma_{\varphi,cal}^2 = \frac{\sigma_y^2}{2m_0^2} . \tag{15}$$

The modulation amplitude can be obtained as

$$m_0 = \frac{1}{2} \sqrt{(\mu_y^{A'} - \mu_y^{B'})^2 + (\mu_y^{A''} - \mu_y^{B''})^2} . \tag{16}$$

The average phase shift μ_φ and the variance σ_φ^2 were investigated at different (simulated[1]) distances. The variances σ_φ^2 and $\sigma_{\varphi,cal}^2$ were compared.

---

[1] See different setups for PMD CamCube and MESA SR3101.

3   Measurement Setups

3.1   Radiometric Measurements

For the measurements of the overall system gain K the sensor was illuminated without a mounted lens by a diffuse disk-shaped light source with a diameter D. An integrating sphere with built in four blue, green and red LEDs served as a light source and was positioned at a distance $d = 8D$ ($f_\# = \frac{d}{D} = 8$ according to the EMVA 1288 standard) from the sensor plane. Fig. 1(left) shows the schematic view of the assembly. The number of photons per area and time interval E_p was varied by changing the LED current. Before, E_p was characterized with an absolute calibrated photodiode PD-9306 (1cm^2) by Gigahertz-Optik placed at the sensor plane. The absolute calibration of E_p is a crucial point in this setup because it depends additionally on the true exposure time of the camera. The current could be varied between 0 mA and 100 mA with at most 4000 steps.

3.2   Spectrometric Measurements

For the measurements of the quantum efficiency $\eta(\lambda)$ the camera was illuminated with light at different wavelengths coming from an arc lamp. A xenon arc lamp was used because of its broadband light with a ZEISS concave diffraction grating as monochromator. The wavelength could be adjusted in the range of 350 nm - 1100 nm with a determined resolution of max. 10 nm. The wavelength calibration was performed with a spectrometer Maya 2000 Pro by Ocean Optics with an internal resolution of less then 1 nm. The irradiance at each wavelength was also calibrated with the photodiode PD-9306 placed at the sensor plane and varied by using different exposure times of the camera.

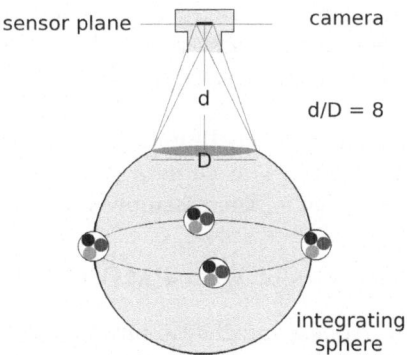

Fig. 1. Measurement setup for radiometric calibration according to EMVA 1288: integrating sphere with built in four blue (470 nm), green (529 nm) and red (629 nm) LEDs

3.3   Phase Shift Measurement with MESA SR3101

The big disadvantage of the usual calibration procedures (e.g. by H. Rapp [6]) is that the irradiance drops inversely proportional to distance square. To avoid this problem, we used two new approaches for calibration of the phase shift. The first idea is, that the light covers always the same distance and the phase shift is performed already in the light source (Fig. 2 (left)). Thus, different distances at a constant irradiance can be simulated.

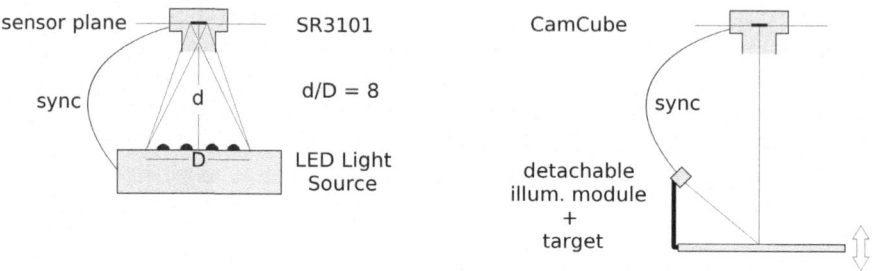

Fig. 2. Setups for phase measurements: MESA SR3101 with PCO light source (left) and PMD CamCube with detachable illumination module and target on the large translation stage (right)

The characterization of the phase information from SR3101 was performed with the PCO Light Source[2] which was developed within the FLICam-Project. It consists of an 4×4 LED array. The intensity can be modulated sinusoidal or rectangular. We used sinusoidal modulation. The modulation frequency can be varied in the range from 400 kHz to 40 MHz and the phase can be shifted relative to the trigger signal from 0 rad to 2π rad with a resolution of 14 bit. A trigger signal coming from the camera was used for the synchronization with the light source. The light source shifts the phase relatively to the trigger signal coming from the camera. Thereby the light source simulates the distance corresponding to this phase shift $\delta\varphi_{\mathrm{LS}}$.

In order to achieve a homogeneous illumination a diffusor was mounted between light source and camera. The camera was also illuminated without a mounted lens. A schematic view of the assembly is shown in Fig. 2 (left).

3.4   Phase Shift Measurement with PMD CamCube

The characterization of the phase information from CamCube at constant object radiance was performed on a large translation stage by keeping the distance of the light source to the target constant. The distance d_{real} between camera and

---

[2] This light source was especially produced by PCO AG for partners in the FLICam project.

target was varied from 0.5 m to 6 m with the accuracy of 1 mm. The planar target with the size of $76 \times 54 \text{cm}^2$ was coated with the optical diffuse material OP.DI.MA from Gigahertz-Optik. The detachable illumination module of the CamCube was mounted at the target as shown in Fig. 2 (right).

4   Results

The calibration procedures described above are performed for two ToF cameras: MESA Imaging SR3101 and PMDTec CamCube. The technical details of these cameras are summarized first, followed by the results of each measurement.

MESA SR3101. The SR3101 by Mesa Imaging is a modified version of SR3000 just without illumination module and filter in front of the sensor. The pixel size is $40 \, \mu\text{m} \times 40 \, \mu\text{m}$. In this version of the camera the variation of the modulation frequency was also possible[3]: 20 MHz and $10 \, \text{MHz}/n$ $(n = 1, 2, 3, \ldots)$. All measurements were done at 20 MHz.

PMD CamCube. The CamCube by PMDTec has a detachable illumination module and no filter in front of the sensor. The pixel size is $45 \, \mu\text{m} \times 45 \, \mu\text{m}$. The modulation frequency is fix at 20 MHz.

Radiometric measurements. Both cameras have a feature to suppress light, which doesn't come from the light source, e.g. not modulated scattered light. For the CamCube this feature is called SBI[4] (suppression of backlight illumination). The SBI initiates at a certain gray value, therefore measurements with continuous illumination described above can be applied in a strictly monotonic increasing part of the sensitivity only (i.e. at small gray values). For the SR3101 this is the case if $\mu_y^A \wedge \mu_y^B < 2200 \, \text{DN}$ and for CamCube if $\mu_y^A < 1900 \, \text{DN}$ $\wedge \mu_y^B < 2000 \, \text{DN}$. Both cameras were measured with the red LED, because their quantum efficiency is the highest in this wavelength range compared to the blue and green LED.

Results of the radiometric measurements are presented in Fig. 3 for both cameras and both potential wells separately.

The responsivity R ist the slope of a linear fit in Fig. 3 (top). The difference between responsivities R^A and R^B is for CamCube larger as for SR3101, the deviation from the linearity however vice versa. Gray values must be summated for the responsivity R of the whole sensor:

$$(\mu_y^A - \mu_{y,\text{dark}}^A) + (\mu_y^B - \mu_{y,\text{dark}}^B) = \mu_p R . \tag{17}$$

This graph is not shown in this paper, because it is here just $R = R^A + R^B$. The gray value in SR3101 shows wiggling behavior as a function of irradiance. This causes major problems to determine the right responsivity.

---

[3] This feature is of interest within the FLICam-Project for fluorescence lifetime imaging.

[4] The acronym SBI will be used in this paper for both cameras to denote the feature described above without the aim to breach the copyright.

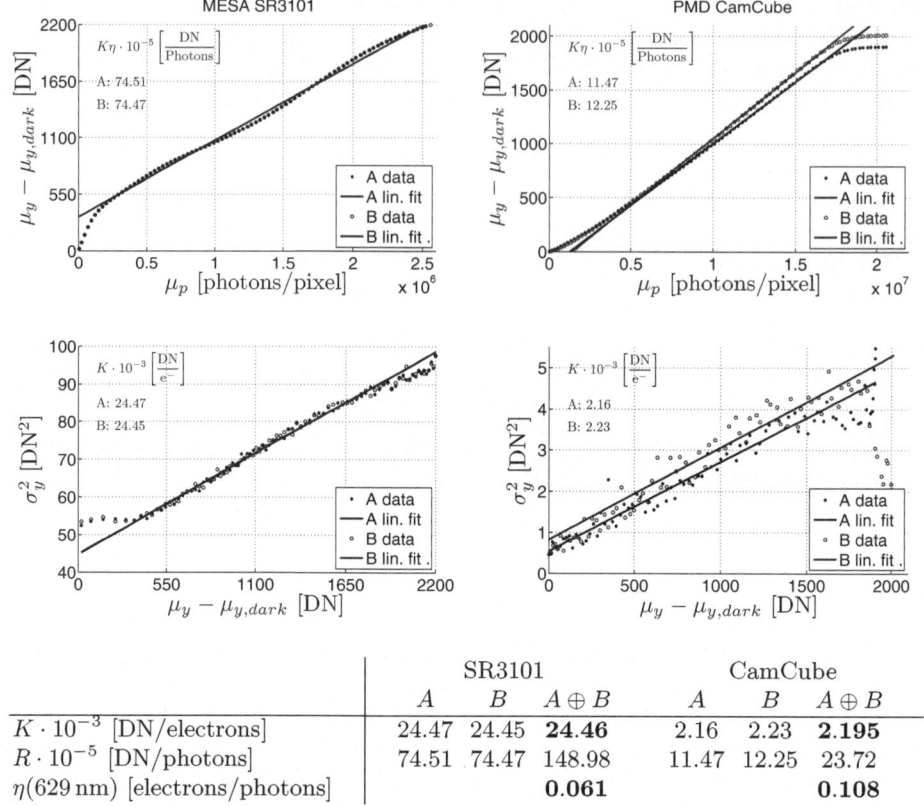

Fig. 3. Results of the radiometric measurements: average gray value μ_y vs. irradiance E_p for one pixel and certain exposure time (top) and variance σ_y^2 vs. average gray value (bottom), for MESA SR3101 (left) and for PMD CamCube (right) with the red LED; for two potential wells A and B, respectively.

The overall system gain K is the slope of the linear fit in Fig. 3 (bottom). K must be independent on the wavelength λ, but because of possible different offsets $\sigma_{y,0}^2$ (see Eq. (7)) the variance was fitted for A and B separately. For CamCube this difference is evident. The overall system gain K is then the average of the two gains

$$K \overset{\triangle}{=} K^{A \oplus B} = \frac{K^A + K^B}{2} . \tag{18}$$

Non-linearity at low irradiances is visible corresponding to the non-linearity in the responsivity graph. The quantum efficiency η was calculated as $\eta = R/K$. The full-well capacity could not be determined because all measurements were done at low irradiances (below the SBI). Thus, the saturation could not occur.

In Fig. 4 the signal-to-noise ratio is presented for both cameras in the measured irradiation range. The $\mathrm{SNR}_{\mathrm{fit}}$ was also calculated for the linear fits of the

Table (from figure):

	SR3101			CamCube		
	A	B	$A \oplus B$	A	B	$A \oplus B$
$K \cdot 10^{-3}$ [DN/electrons]	24.47	24.45	**24.46**	2.16	2.23	**2.195**
$R \cdot 10^{-5}$ [DN/photons]	74.51	74.47	148.98	11.47	12.25	23.72
$\eta(629\,\mathrm{nm})$ [electrons/photons]			**0.061**			**0.108**

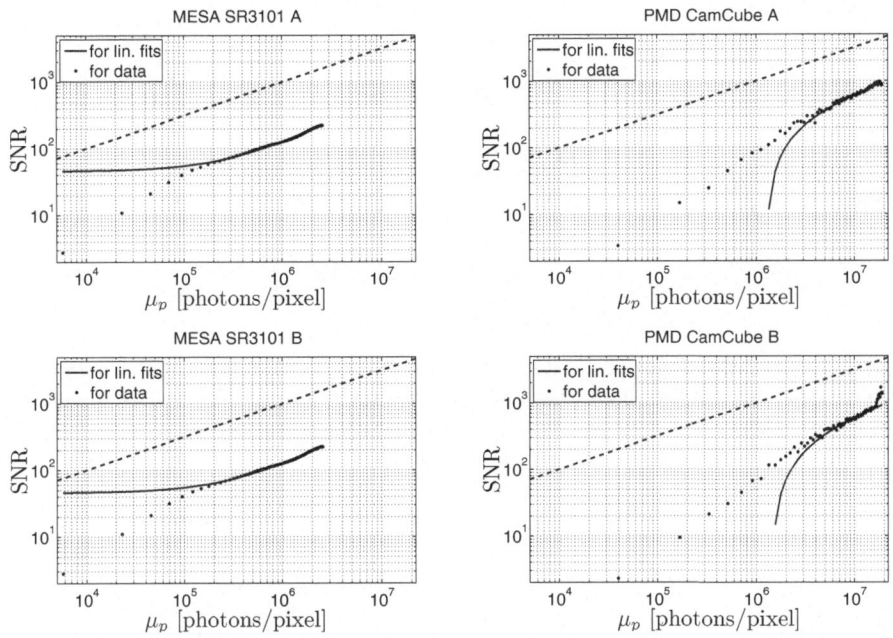

Fig. 4. Results of the radiometric measurements: Signal-to-noise ratio for MESA SR3101 (left) and for PMD CamCube (right) and for two potential wells A (top) and B (bottom), respectively; The dashed line is the theoretical limit of $\sqrt{\mu_p}$

sensitivity and noise within the given irradiation ranges. The dashed line is the theoretical limit of $\sqrt{\mu_p}$ for an ideal sensor with only shot noise and a quantum efficiency of one. For CamCube the decline in $\mathrm{SNR}_{\mathrm{fit}}$ for small irradiances is caused by negative offset in the linear fit, see Fig. 3 (top right). Both cameras seem to have nearly the same SNR in the range of $2 \cdot 10^6$ photons/pixel. At lower irradiances SNR for SR3101 is higher.

Quantum efficiency $\eta(\lambda)$ measurements. The quantum efficiency η was measured at $\lambda = 350\,\mathrm{nm}$ to $1050\,\mathrm{nm}$ in 530 steps. The results for both cameras are presented in Fig. 5.

Because of the non-linearity of the sensor the quantum efficiency for SR3101 is very erratic. As mentioned above non-linearity in the sensitivity of the sensor causes problems during the determination of responsivity. Wiggling in the responsivity (Fig. 3) implicates very erratic determination of the slope and thus of η. Because of the very small scale (10^{-5}) small changes in the slope already cause clear changes in the quantum efficiency.

On the other hand the spectrum of the Xe arc-lamp shows very high peaks in IR range. Thus, the quantum efficiency can be measured only piecewise with adapted exposure time. Therefore, the non-linearities play again an important role.

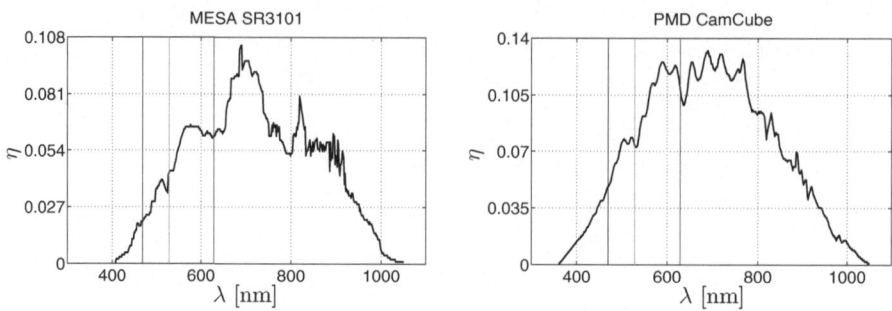

Fig. 5. Results of the quantum efficiency measurements for MESA SR3101 (left) and PMD CamCube (right). The blue (470 nm), green (529 nm) and red (629 nm) lines represent wavelengths of three LEDs built in the integrating sphere.

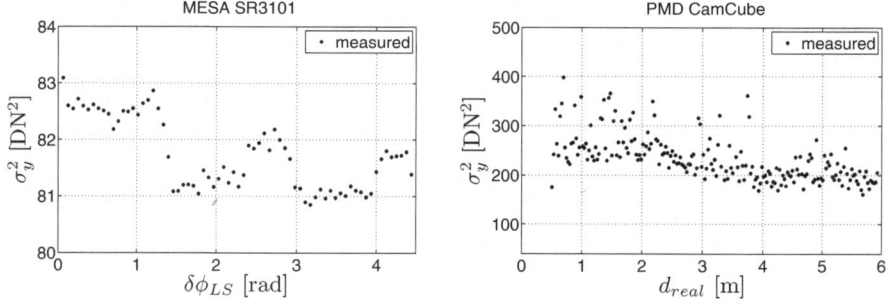

Fig. 6. Variance of gray values σ_y^2 for SR3101 and CamCube

The fill factor of the CamCube sensor is about 26% (personal communication T. Ringbeck from PMDTec), but the maximum quantum efficiency is only about a half of it, which is a quite good value. Micro lenses would be able to increase the effective quantum efficiency considerably. The fill factor of the SR3101 sensor was unknown.

Phase shift measurements. The variance of the phase shift σ_φ^2 was investigated for SR3101 with the setup shown in Fig. 2 (left). The relative phase shift done by the light source $\delta\varphi_{\mathrm{LS}}$ was varied in the range $0-2\pi$ rad with 100 steps.

The variance of the phase shift for CamCube was investigated with the setup shown in Fig. 2 (right). The distance between camera and target was varied between 0.5 m and 6 m with 200 steps. In Fig. 6 the variance of gray values σ_y^2 is shown for both cameras.

The average phase shift μ_φ was calculated with Eq. (5). The variance of the phase shift σ_φ^2 was calculated with Eq. (12) and compared with the theoretical value $\sigma_{\varphi,cal}^2$ (see Eq. (15)).

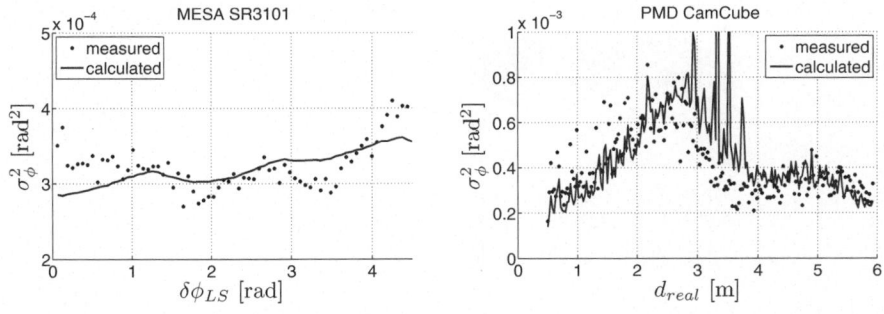

Fig. 7. Variance of the phase shift σ_φ^2 compared with the calculation $\sigma_{\varphi,cal}^2$ for SR3101 and CamCube

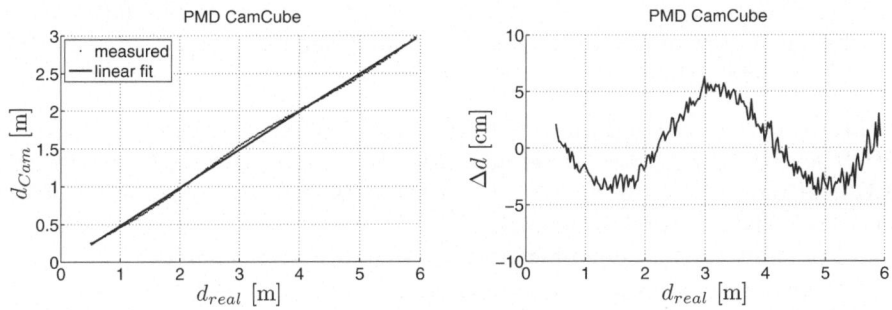

Fig. 8. The distance measured by CamCube d_{Cam} (left) and the wiggling error Δd (right) vs. the real distance d_{real} on the large translation stage

The agreement is quite good. The small differences for SR3101 probably trace back to the non-linearity of the responsibility of the sensor. Outliers for Cam-Cube between 3 m and 4 m are caused by large variance of the gray values, which are located outside the plot boundaries. The reason are probably short-term changes of the ambient light.

Since the description of systematic errors in the distance measurement is not the goal of this paper, just the wiggling is shown for the CamCube in Fig. 8. It should demonstrate the advantage of the used setup with constant illumination.

The measured distance d_{Cam} [m] is just one half of the real distance d_{real}[m] on the large translation stage, because the light covers just half the distance assumed by the camera for calculations.

Nonuniformities in the data. According to the EMVA 1288 there are three types of spatial distributions of gray values wich were not investigated in this paper:

– Dark Signal Nonuniformity, DSNU: the distribution of the dark signal, which is a part of the dark value $\mu_{y,\mathrm{dark}}$ and well-known as Fixed-Pattern-Noise;

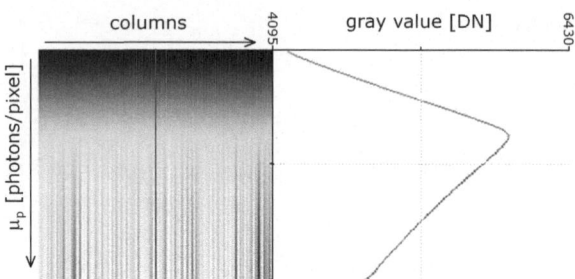

Fig. 9. Gray values for one row in the image sensor of PMD CamCube at different irradiances (left) and exemplarily for one pixel (right)

- Photoresponse Nonuniformity PRNU: basically the distribution of the overall system gain K;
- Dark Current Nonuniformity, DCNU: the distribution of the dark current, which is a part of the dark value $\mu_{y,\mathrm{dark}}$ and depends on the exposure time and ambient temperature.

Qualitatively, the nonuniformities DSNU and PRNU are typical distinct for CMOS sensors. For SR3101 these quantities vary more than for CamCube. For SR3101 at modulations frequencies lower than 1 MHz synchronization problems cause the DSNU to vary with time.

Additionally to the nonuniformities presented above differences in the SBI for each pixel are visible. This is demonstrated for one row of the CamCube sensor in Fig. 9. The nonuniformity in gray values after the initiation of SBI is observable. This effects of the SBI are modeled in detail in [11].

5   Conclusion and Future Work

This paper proposed to extend the EMVA 1288 standard for lock-in pixels. The applicability of procedures described in the standard could be investigated and confirmed with two ToF cameras: PMDTec CamCube and MESA SR3101. The characterization of radiometrical and electrometrical sensitivities and of the noise could be transferred to the lock-in pixels. An adaption of quantum efficiency definition to the split illumination time was necessary. Because of the suppression of ambient light in both cameras, the sensitivity could be measured in the range of only small gray values. Therefore, it was not possible to measure the full-well capacities. The distance/phase shift and its variance were measured with both cameras but in different setups. The noise in the phase measurement could be related to the noise in the intensity. A good agreement was found for both cameras.

Nonuniformities of the signal included in the EMVA 1288 standard were not investigated yet. A nonuniformity of the SBI was found additionally to the quantities described in the standard. Also the demodulation contrast was not measured yet, even though it's one of the most important quantities.

Before the definition of the extended version of the EMVA 1288 standard the general measurement conditions must be clarified, e.g. the target material used for the distance measurements. Important figures and/or specific sensor characteristic quantities describing the performance of the lock-in sensor with respect to distance measurement must be defined.

Acknowledgements

We thank PCO AG, especially Robert Franke, for the LED Light Source and support in terms of electronics, Mirko Schmidt for programming the camera driver and for great support during the measurements. We gratefully acknowledge financial support by the BMBF (German Federal Ministry of Education and Research), through projects FLICAM and Lynkeus.

References

1. EMVA: EMVA Standard 1288, Standard for Characterization of Image Sensors and Cameras. European Machine Vision Association (2009)
2. Schmidt, M.: How to perform camera measurements according to the EMVA 1288 Standard. Talk on Industrial Vision Days, Vision Stuttgart (2008)
3. Lange, R.: 3D time-of-flight Distance Measurement with Custom Solid-State Image Sensors in CMOS/CCD-Technology. University of Siegen (2000)
4. Kahlmann, T., Ingensand, H.: Calibration and improvements of the high-resolution range-imaging camera SwissRanger. In: Proc. SPIE, vol. 5665, pp. 144–155 (2005)
5. Lindner, M., Kolb, A.: Lateral and Depth Calibration of PMD-Distance Sensors. In: Bebis, G., Boyle, R., Parvin, B., Koracin, D., Remagnino, P., Nefian, A., Meenakshisundaram, G., Pascucci, V., Zara, J., Molineros, J., Theisel, H., Malzbender, T. (eds.) ISVC 2006. LNCS, vol. 4292, pp. 524–533. Springer, Heidelberg (2006)
6. Rapp, H.: Experimental and Theoretical Investigation of Correlating ToF-Camera Systems. University of Heidelberg (2007)
7. Hempel, M.: Validierung der Genauigkeit und des Einsatzpotentials einer distanzmessenden Kamera. Diplomarbeit, TU Dresden (2007)
8. Frank, M., Plaue, M., Rapp, H., Köthe, K., Jaehne, B., Hamprecht, F.A.: Theoretical and experimental error analysis of continuous-wave time-of-flight range cameras. Optical Eng. 48, 13602 (2009)
9. Keller, M., Kolb, A.: Real-time Simulation of time-of-flight Sensors. J. Simulation Practice and Theory 17, 967–978 (2009)
10. Peters, V., Loffeld, O., Hartmann, K., Knedlik, S.: Modeling and Bistatic Simulation of a High Resolution 3D PMD-Camera. In: EUROSIM 2007 (6th EUROSIM Congress on Modelling and Simulation), Ljubljana, Slovenia (2007)
11. Schmidt, M.: A physical model of time-of-flight 3D imaging systems, including suppression of ambient light, Dynamic 3D Imaging. In: Koch, R., Kolb, A. (eds.) Dyn3D 2009. LNCS, vol. 5742, pp. 1–15. Springer, Heidelberg (2009)
12. Janesick, J.R.: CCD characterization using the photon transfer technique. In: Prettyjohns, K., Derenlak, E. (eds.) Solid State Imaging Arrays. SPIE Proc., vol. 570, pp. 7–19 (1985)
13. Janesick, J.R.: Photon Transfer. SPIE Press, San Jose (2007)

Datastructures for Capturing Dynamic Scenes with a Time-of-Flight Camera⋆

Ingo Schiller and Reinhard Koch

Institute of Computer Science
Christian-Albrechts-University (CAU)
24098 Kiel, Germany
{ischiller,rk}@mip.informatik.uni-kiel.de

Abstract. To capture 3D dynamic scenes, a suitable 3D data structure is needed that can represent the dynamic scene content. In this contribution we analyse and evaluate different data structures for capturing time-varying depth and color data of a dynamic scene obtained with Time-of-Flight and color cameras. The comparison of depth-panoramas, layered depth images and volumetric structures shows that a volumetric octree is best suited to fuse time-varying 3D scene data. We exploit the octree data fusion capabilities for different application scenarios, like 3D environment building, volumetric object reconstruction, and geometric interaction.

1 Introduction

Recently, Time-of-Flight- (ToF) cameras have made significant progress as alternative to range measurement techniques such as laser scanning, structured-light, or stereo-camera-systems to name the most popular ones. The ToF-cameras which are available today offer a resolution of up to 204×204 pixel and framerates up to 25 Hz. This results in a large amount of data, especially if every pixel is seen as an independent measurement of a threedimensional point (one second of recording produces 1.040.400 points at 25 Hz). ToF-cameras can be used in different scenarios, in which the camera may be static or moving and the observed scene is either static or parts of it are dynamic. Each scenario produces different requirements on the representation of the data. A suitable 3D datastructure for holding ToF-camera data therefor has to provide the following properties:

1. Spatiotemporal expandability
2. Inherent data fusion of multiple measurements
3. Representation of geometry and occlusions
4. Possibility to represent dynamic scene content

⋆ This work was partially supported by the German Research Foundation (DFG), KO-2044/3-2 and the Project 3D4YOU, Grant 215075 of the ICT (Information and Communication Technologies) Work Programme of the EU's 7th Framework program.

R. Koch and A. Kolb (Eds.): Dyn3D 2009, LNCS 5742, pp. 42–57, 2009.

5. Hold intensity and depth information simultaneously
6. Hold different data for different spatial viewpoints and points in time
7. Generation of a dense surface
8. Runtime and storage efficiency

In this contribution we will discuss several datastructures for holding ToF-data and select the most suitable one. Various applications are presented which make use of the characteristics of the selected datastructure, such as environment reconstruction (section 5.1), reconstruction of the outer hull of objects (section 5.2), Space-Time-Video (section 5.3) and geometric interaction (section 5.4).

2   Data Capture with ToF- and Color Cameras

2.1   ToF Measurement Principle

The Time-of-Flight-camera is a 2.5D camera which delivers dense depth images with up to 25 frames per second. The camera actively illuminates the scene by sending out incoherently modulated light with a modulation frequency between 20 and 31MHz. This light is reflected by 3D scene points and received by the image sensor, a semiconductor structure based on CCD-/ CMOS-technology ([1],[2]), of the ToF-camera. The received and the emitted signal are correlated in the camera's image sensor and the phase shift between the two signals is determined in every pixel. This phase shift depends on the time-of-flight and the distance of the scene can therefor be calculated from this phase shift. With the used modulation frequency of 20MHz the non-ambiguity range of the ToF-camera is 7.5 meters. Since the resolution of the phase difference measurement is independent from distance, the achievable depth resolution is independent from scene depth. This is in contrast to stereo triangulation where depth accuracy is proportional to inverse depth. However the signal-to-noise ratio decreases with increasing distance to the camera due to the quadratic light intensity fall off.

2.2   Camera Projection Model

The ToF-camera uses traditional optics and a standard lens and the camera geometry is therefore characterized by the projection matrix (cf. [3, p.143]):

$$P = KR^{\mathsf{T}}(I_3| - C) \tag{1}$$

with K the camera matrix, R the rotation matrix, I a 3×3 identity matrix and C the camera center. The projection matrix maps homogeneous 3D points $\mathbf{X}$ to homogeneous image coordinates $\mathbf{x}$ according to:

$$\lambda(\mathbf{x})\mathbf{x} = P\mathbf{X} \tag{2}$$

Since the depth $\lambda(\mathbf{x})$ of a pixel $\mathbf{x}$ in an image is measured by the ToF-camera, the homogenous 3D point $\mathbf{X}$ can be calculated as:

$$\mathbf{X} = \lambda(\mathbf{x})RK^{-1}\mathbf{x} + C \qquad (3)$$

So for every pixel in a depth image a 3D point can be generated.

2.3   Data Fusion with Color Cameras

For many applications it is necessary to combine the ToF-camera with a standard color camera. For example environment reconstruction in depth and color, registration by feature matching, matting in depth and intensity and mixed reality applications. A good overview of the state-of-the-art of algorithms and applications with ToF-cameras can be found in [4]. In most cases the color camera has a much higher resolution than the ToF-camera, e.g. 1600×1200 compared to 176×144 pixel. For metric measurements the ToF-camera has to be calibrated, which means that the elements of the camera matrix K, such as focal length, principle point and radial and tangential distortion have to be estimated. To use the ToF-camera together with the color camera both have to be calibrated and the relative translation and rotation between the cameras have to be known. Furthermore other error sources distort the depth measurement of the ToF-camera. These are described in detail in [5] and [6]. The used calibration approach relies on the method described in [7] and [8] which includes depth error calibration and intrinsic and extrinsic camera calibration.

For applications which require high resolution depth maps it is necessary to transfer the depth image of the ToF-camera into the color camera. We use the efficient way to mesh the depth image on the graphics card and to render it with the internal and external camera parameters of the intensity camera as described

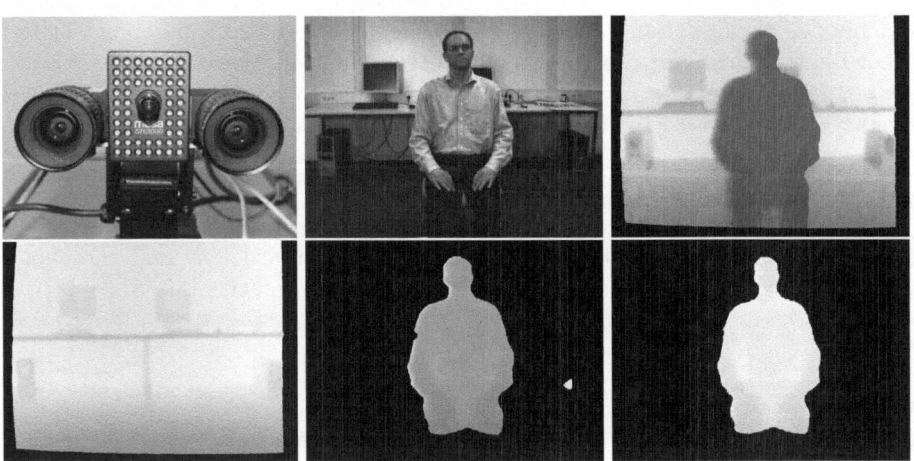

Fig. 1. Upper row: Camera rig consisting of two color cameras and a Swissranger3000 ToF-camera, current color image (1024x768 px) and current depth image (176x144px), warped into the view of the color camera. Bottom row: Warped background depth image, thresholded current depth image and current depth image after erosion.

in [9]. This way a 2D-3D video stream is obtained, where every pixel holds color and depth values. In figure 1 the used camera rig, a color image and a warped depth image are shown in the upper row.

Note that the background and foreground are connected on the left side of the person in the rightmost image which is obviously wrong. Reduction of this error and simultaneous foreground segmentation is done by thresholding. The lower row of figure 1 shows a warped depth image of the background on the left. The current depth image is thresholded against the background image which almost completely removes the background. But still the false connection between foreground and background is present at the contour of the person. This is further addressed by applying a shrinking of the silhouette by erosion.

3   Data Representation Review

In this section we give an overview of the existing representations of ToF-camera data. The advantages and disadvantages of each representation are discussed.

1. **Unstructured 3D Point Cloud**

   Applying the projection matrix of the ToF-camera and Eq. (3), the delivered depth image can be transformed into an unstructured 3D point cloud. This is for example used in [10]. Considering a natural scene, the noise present in the measurements and the given limited accuracy of the ToF-camera, every scene point will result in different measurements which will create multiple 3D points for the same scene point. This results in a large amount of points which do not have any neighboring relations. Additionally closed surfaces are not represented as such but split up into a number of independent points. Furthermore, there is no updating of already measured scene points. Averaging over time or measuring points multiple times can increase robustness towards outliers. Point clouds do not offer such possibilities. Constructing a (partially-) closed surface from an unstructured point cloud is a very demanding task and can often not be solved to satisfaction.

2. **2.5D Panoramic Image (Depth-color panorama)**

   In [9] and [11] the representation of the data is chosen as a 2.5 dimensional panoramic image that encodes color and depth in a 2D representation, for example in planar, cylindrical or spherical coordinates. With panoramic image representation, measurements which are taken with a rotating camera head can be fused. Multiple measurements of the same points are fused by averaging in the image, making the result more robust towards outliers. Unfortunately, this representation has some disadvantages. It is by nature only a 2.5 dimensional representation of a threedimensional scene and occlusions can not be represented. Furthermore, dynamics are not realizable in an efficient way. An advantage, however, is that from a panoramic depth image it is easy to construct a (partially-) closed surface representation, such as a triangle mesh using delaunay triangulation, which can be rendered efficiently on the GPU. An example of panoramic representation is discussed in section 5.1.

3. **LDI, 2.5D Layered Depth Image**
   Shade et. al. [12] introduced the layered depth images. A layered depth image is an image in which at every position multiple depth and color measurements are stored corresponding to the line of sight through that pixel. LDIs were developed for image-based-rendering, which describes an approach to generate new interpolated views of a scene. Thus LDIs are capable of representing occlusions or dynamics, but not both at the same time. LDIs are constructed for a distinct camera position. Generating LDIs includes the warping of all depth images to that camera position. LDIs can be viewed as generalization of depth panoramas to multiple occlusion layers. Rendering a scene from a different view requires to perform the incremental warping procedure (cf.[12]).

4. **3D Volumetric Representation**
   Volumetric models divide the space into volumetric entities of a given size. The most widely known and used model is the Voxel representation as used in [13]. Several tree structures have been discussed (e.g. in [14],[15]), which are favorable due to their hierarchical nature.

The representation of multiple measurements as point cloud, panoramic image or LDI is not optimal. The main disadvantages are either the missing fusion of measurements and the lack of neighboring relations, or the missing possibility to represent occluded objects and dynamic content. The obvious step towards an optimal representation is to use a true volumetric representation of the scene.

4   The Octree Datastructure

We conclude from section 3 that a volumetric representation of data is needed to provide all required characteristics. The octree representation [16] combines the advantages of a volumetric model with a hierarchical data structure. Further advantages are ease of concept and implementation through recursion, storage efficiency and flexibility concerning volume content. We will therefor introduce the used octree structure, show how the fusion of measurements is done and derive why it is in our opinion the optimal data structure to represent ToF-camera data.

4.1   Building the Octree Structure

The octree data structure is an oriented graph structure which represents a part of threedimensional space. It is a recursive datastructure in which every octree has eight octree children and one octree father. Each octree node consists of a position in 3D, a size and an octree element, representing the information about space in this volume element.

During initialization the octree holds only one element with a certain size containing the bounding volume to be modeled. Before adding measurements the spatial resolution of the octree has to be defined. This definition is made from knowledge we have about the used ToF-camera. For current ToF-cameras the manufacturers promote a repeatability between 5 and 200mm. So this can be used as guidance to select the minimal octree cell size, which we normally set to $25mm^3$. To add measurements the algorithm starts at the top node by checking if the size of the node to add is bigger than half the size of this node. If this is the case a leaf is reached and the node is inserted here, if not the algorithm calculates to which child node the node to add belongs. If the child node does not already exist it is created and this node is used in further branching.

4.2   Measurement Fusion

Depending on the requirements, different elements can be inserted in the octree branches. This reaches from simple uncolored 3D points over colored points with normals up to small oriented surface patches with texture. Using natural scenes and lighting, objects can look different when viewed from different angles, this can also be included as well as different appearances depending on daytime or other factors. As every child of an octree is an octree itself, subtrees can easily be added to the current scene. In contrast to simple point clouds where in general no measurement fusion is possible, the volumetric representation of the octree allows to fuse the measurements while adding them. In our experiments we use simple colored 3D points with an additional radius component as octree elements and to fuse multiple measurements we average the new and the already existing position and color.

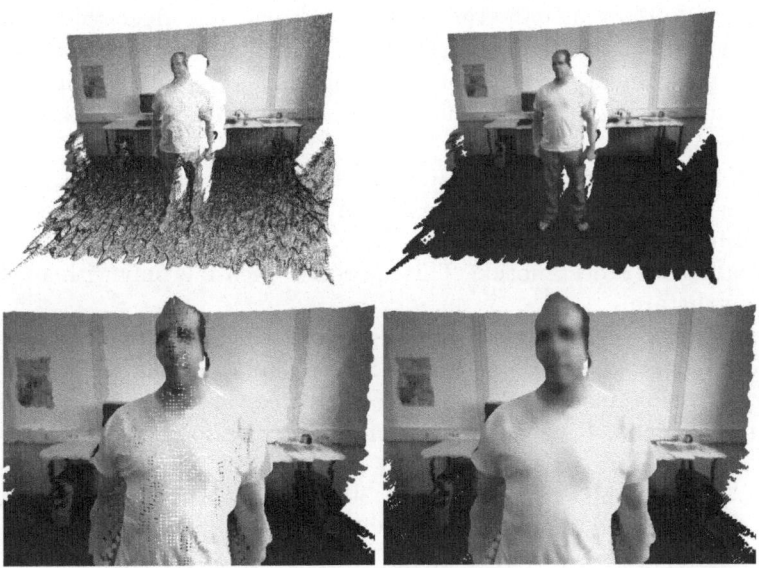

Fig. 2. Comparison of point rendering (left) and point splatting (right)

4.3   Rendering / Surface Generation

Rendering octrees is fast and straightforward. It consists of collecting all active nodes in the octree and rendering them. How the octree elements are rendered is mainly due to the intended usage. If only a sparse point cloud is needed, the octree can be rendered as points with color and a certain size. This is shown in figure 2 in the left column. If a closed surface is needed, e.g for depth testing many existing approaches are usable. For example point splatting (cf.[17]) is a feasible method exploiting GPU shader language. Examples of point splat rendering can be found in figure 2 on the right. Not only point based rendering methods are applicable, for example in [18] Samet shows how raytracing can be efficiently performed using octrees.

5   Analysis and Exploitation of Suitable Data Structures

In this section we will analyse and exploit the data structures for their use in different application scenarios. We look at typical scenarios, like 3D environment scanning from a rotating camera head (looking inward-out), modeling a volumetric object by rotating it in front of the camera head (looking outward-in), and the analysis and capture of dynamically moving objects (persons) over time.

Environment model generation is the task of capturing and aligning multiple depth and intensity measurements, and of forming a consistent threedimensional model from the images taken by a rotating camera head. This is a common application, in which ToF-cameras can be used due to the instantaneous availability of depth information and the predominance of ToF-cameras compared to stereo algorithms in untextured areas. We will introduce the environment model construction of indoor rooms using texture and depth information. Focus is laid on the comparison of 2.5D panoramic image representation and hierarchical volumetric representation of the data. Both representations are feasible in this scenario.

In section 5.2, we will discuss volumetric modeling of rigid objects. The camera head observes a person which is rotating on a swivel chair, and all measurements are fused into a consistent volumetric model while simultaneously tracking and compensating the object motion. This approach is feasible only with a volumetric 3D data structure.

In section 5.3, we discuss a dynamic scenario which fully exploits the benefits of the octree with Space-Time-Video. Space-Time-Video is the recording and playback of threedimensional video, allowing time-varying objects in the scene.

The last section 5.4 is devoted to geometric interaction and collision detection using the octree data structure.

In our experiments, a SwissRanger 3000 ToF-camera with a resolution of 176×144 pixel is used together with a CCD-camera with 1024×768 pixel. For the environment model construction, an integration time of 160ms (6 fps) of the ToF-camera is used to reduce noise in the images, whereas in the other applications an integration time of 80ms (12 fps) is used to reduce motion artefacts.

5.1   Environment Model Reconstruction with Pan-Tilt-Unit

In this experiment, we compare 2.5D panoramic image and octree usage. A camera head with ToF and color camera is mounted on a pan-tilt-head, scanning the environment. We compare a representation as 2.5D cylindrical depth-panorama with a 3D octree reconstruction.

The 2.5D panoramic image has a resolution of 2000×1000 pixel. The images taken from the two cameras are fused and stitched into the 2.5D panoramic images which are shown in figure 3.

Fig. 3. 2.5D panoramic images for texture and depth

The depth-color panorama is converted into a trianglemesh using the delaunay triangulation, and a threedimensional surface representation is generated, as seen in figure 4. As the camera head is rotating in the middle of the room, the disadvantages of the 2.5D representation are not important in this case.

The same input images of the camera head were used to construct the octree model shown in figure 5 where a depth resolution of 25mm has been chosen and the measurements have been fused in the octree cells while being added to the octree.

A visual comparison of both models after rendering shows that the surface mesh is somehow smoother than the octree rendering, but both models are comparable in visual quality. The octree processing can also be compared to simple

Fig. 4. Triangle mesh generated from 2.5D panoramic image. Top Left: View of complete environment model. Others: Three views of the interior.

Fig. 5. Environment model as octree, fused from 115 images, cellsize $25mm^3$. A slight degradation due to point rendering is visible.

3D point cloud construction and rendering. Table 1 compares the performance of the octree for the reconstruction of the indoor scene for different cell sizes and simple 3D point cloud processing. The scene has been constructed for different cell sizes. Insertion of one image with 176×144 pixel and merging it with existing content is performed in 44ms ($100mm^3$) to 69ms ($25mm^3$). This includes the computation of 3D points which is also done for the point cloud. So the real traversal in the tree and the merging takes between 6ms ($100mm^3$) and 30ms ($25mm^3$). Traversing the tree and collecting all valid elements takes between 11ms and 192ms and rendering all points takes 2ms to 32ms. From these numbers it is observable that at a cell size of $50mm^3$ the octree rendering is compareable to pure point cloud rendering concerning speed. The tests have been carried out using an Intel Core2 CPU 6600 @ 2.40GHz with 4GB RAM and a NVidia GeForce 8800 GTS GPU. The indoor reconstruction scenario shows the suitability of the octree datastructure to represent ToF-measurements.

Besides the performance the storage efficiency is a crucial point concerning the choice for a data representation. Comparing image based representations such as 2.5D panoramic images or LDIs and 3D representations it is obvious that volumetric representations will require more storage because instead of one value for depth, three values for the 3D position are saved. The storage comparison is given in table 2.

Table 1. Comparison of octree cell size and 3D point cloud performance for the environment model. See text for details.

	Pointcloud	Octree $25mm^3$	Octree $50mm^3$	Octree $100mm^3$
Nr. of Elements	2 744 772	1 617 140	470 230	98 102
Insert in Octree	38,48 ms	68,80 ms	51,46 ms	44,38 ms
Collect Elements	-	191,55 ms	51,32 ms	10,80 ms
Render Elements	59,83 ms	32,30 ms	9,69 ms	2,15 ms

Table 2. Comparison of storage requirements for 2.5D panoramic image, panoramic image as triangle mesh, pointcloud and octree. Nr.of Elements denotes the number of points or triangles, The row denoted "In RAM" means the real storage consumption for loading the data into memory and the row "On disk" is the file size for saving the data to disk as images (binary), VRML or octree (ASCII).

	Panorama	Mesh (VRML)	Pointcloud	Octree $25mm^3$	Octree $50mm^3$	Octree $100mm^3$
Nr. of Elements	2 000 000	2 305 030	2 744 772	1 617 140	470 230	98 102
In RAM[MB]	-	1 085.85	115.55	735.26	200.02	41.68
On disk[MB]	13.67	136.37	174.18	344.64	100.22	21.19

Two images are used for the panoramic image, one holding the depth values and one for the intensity information. For depth float values and for intensity RGB values in unsigned char with one Byte per channel are used, which results in 4 Bytes for a float and 3 Bytes for intensity for every pixel. The generated triangle mesh is bigger as for every pixel the 3D position and connection information is saved as well. The texture is saved as an image as above and projected on the geometry. The storage requirement is dependent on the texture resolution. In this case full resolution (2000×1000) has been used and the storage usage is over 1 GB. Lower numbers are observable for the pointcloud as only the 3D position and intensity values are saved. For the octree the 3D position of the cell, the 3D position of the point, four cornerpoints of each cell, the cell size and the pointers which connect tree items are necessary. This produces a far higher memory usage than the representation as pointcloud, but for an octree with a cellsize of $50mm^3$ the memory consumption is only moderately increased. This shows that large data sets which cover large environments, in this case approximately $5 \times 7 \times 3$ meters and larger, are manageable with a resolution of $25mm^3$ and smaller using octrees.

5.2 Rigid Object Modeling

In this scenario, we need a true 3D data structure to model closed surfaces. We want to model the 3D surface geometry of a person by estimating the movement of the person and by fusing the measurements of ToF- and CCD-camera from all views into a single model, while the person is turning on a swivel chair in front of the cameras. Due to self-occlusions of the object, it is impossible to use 2.5D panoramic images, whereas the volumetric representation is capable of representing the data and fusing the multiple measurements. In total 64 images, one of it is shown in figure 1, have been fused. In this scenario we assume that the object itself is rigid but may rotate in front of the camera, or the camera may move around the rigid object.

The motion estimation is based on the tracking of corner features and pose estimation on 2D-3D correspondences. The foreground object is segmented by depth keying from background, since the depth gives an easy cue to object segmentation here. In the first image of the sequence, intensity corners are detected (we use the features described in [19]) and these are tracked in the subsequent images.

For every 2D feature $\mathbf{x}$ the distance $\lambda(\mathbf{x})$ to the camera center is taken from the warped and filtered depth image if the feature is on the segmented foreground. Using equation 3, a homogenous 3D point $\mathbf{X}$ is generated for every feature on the segmented person. From these correspondences, the camera pose is estimated using the standard camera pose estimation scheme DLT (cf. [3] p.173 and p.73).

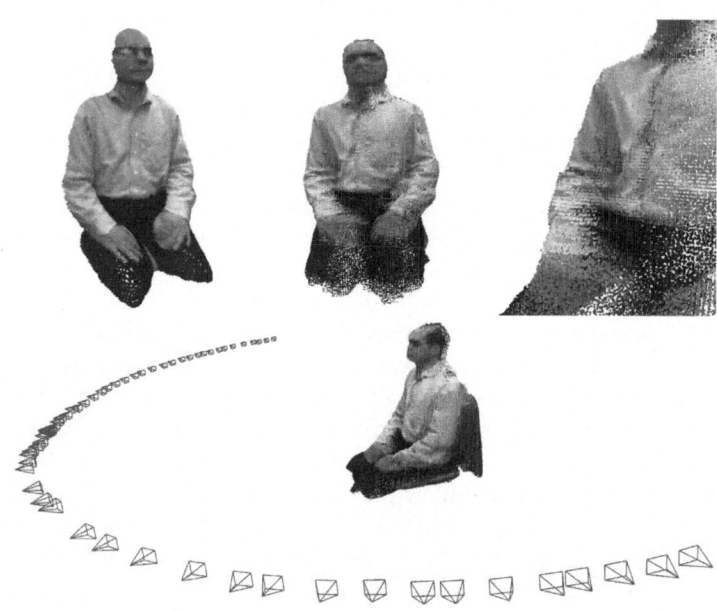

Fig. 6. Person model as octree, fused from multiple image pairs. Top left: octree fused from 5 images. Right: person model fused from 64 image pairs. The missregistration is mainly due to local motion of the person during recording. Voxel cells were rendered as point cloud without splatting. Bottom: Fused person model with camera poses drawn as pyramids.

After estimating the object pose for every image, the depth and intensity elements are added into an octree. Figure 6 shows the resulting fused model. The top left image shows a result after five images, the other images show the integration of 64 images. Note that the person swayed a little during recording, which leads to some errors in the fused model. The bottom image shows the reconstructed person and the estimated camera position as red pyramids, as seen from the rigid object coordinate system. The fused model still contains some errors, but the objective is to show the advantages of the volumetric representation compared to other data structures.

5.3   Space-Time-Video

The above scenarios handle rigid objects only. For truly dynamic scenes, local object deformation must be considered. One way to handle a dynamic scene is Space-Time-Video, that does not only encode the threedimensional geometry of the scene but records also the change in time as 4D representation. The replay of Space-Time-Video should allow to separate 3D space and time, which allows to render the scene freely from any view point at any time. This requires that for every time step the full 3D geometry is known. Approaches to Space-Time-Video include for example the image-based rendering approaches as in [20] and [21] but these lack the possibility to represent non-rigid objects.

An approach in which human motion is recorded and actions are represented as Space-Time Shapes can be found in [22]. There, actors are segmented using silhouette information, and shapes are recorded over time. These shapes are stored in a volumetric representation and classification of motions is applied. As ToF-cameras are capable of recording real-time depth image sequences, these cameras are well suited for recording Space-Time Video. To record and replay Space-Time Video, we extend the octree datastructure in time and store in every cell the time at which it is visible. In this example, we show the scene with and without background. Removal of the background means as well, that the geometry in the back of the person is obviously incomplete. This restricts our camera movements, which however can be solved by using additional ToF-cameras that observe the scene from different viewpoints. When adding a new image to the octree datastructure, the current image number is used as a timecode. If the volume element is already occupied, the geometric information is calculated as an average of old and new measurement and the timestamps at which the element is visible are updated.

As we have a volumetric representation of our scene at hand, we can freely move the camera around and render our octree by selecting only those cells which were visible at a given time stamp. Figure 7 show examples of such a Space-Time-Video sequence, including camera movement. A person is standing in front of the camera head, swinging the arms. The left image shows all octree elements which have been filled during the animation sequence which consists of 375 images. The second image shows a selected frame of the sequence. Image three and four show one moment in time, recorded by the ToF-Camera and rendered from different views. By including the object in the environment octree model, the background

Fig. 7. Space-Time Video: All elements in the octree (left), selected time instance of the animation sequence (second), rendered views with background (right).

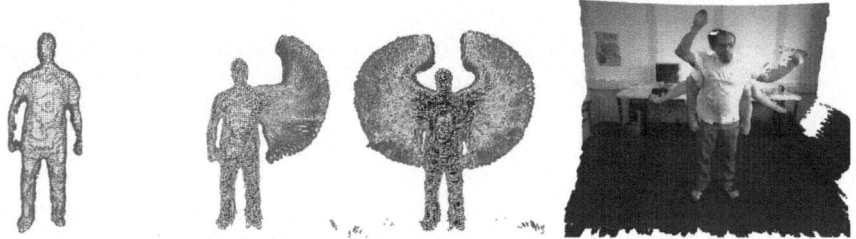

Fig. 8. Left: Accumulated hitcount for video sequence over time, color coded: red (< 60) - black ($>= 360$). From left to right: Frame 1; Integration of frame 1 to 120; integration of frame 1 to 375. Right: Rendering by selection of 3 time frames of the Space-Time Video.

can be filled as well. An example of model-background integration can be seen in fig. 9, top.

Figure 8 shows the colorcoded accumulated hitcounts for the above sequence. The number of times an octree volume element is hit up to the current time frame is saved and analyzed. This provides information about which parts of the scene are static and which are dynamic. In the above sequence, the arms of the person are moving and therefore distinguished from the body. 4D Space-Time representation also allows to combine 3D information which has been taken at different points in time. This is the selection of different time slices and the combination in one 3D shape. An example is shown on the right of figure 8 in which three points in time have been selected, forming a person with six arms.

5.4   Interaction and Collision Detection

The octree representation allows for fast 3D occupancy detection in the scene, which can be exploited for collision detection between real and virtual objects. Collision detection uses the collision check with any occupied node of the octree. This way collision checks with other threedimensional objects can be performed.

Fig. 9. Top: Octree of the environment. Shown are the colored (left) and the depth-coded (right) voxels. Bottom: two frames of the animation sequence.

To show the capabilities of this representation, we present an example of mixing and interaction of real dynamic content and virtual content. First we build an environment model as described in section 5.1 and represent it as an octree. Here we process a volume of $8 \times 8 \times 8m^3$ with a minimum element size of $(50mm)^3$, which is also the bounding box for collision detection. In the interaction phase, we use this environment model together with dynamic object modeling from the ToF-camera to detect dynamic objects in the scene. Depth-keying is applied to segment the object from the background (see [9]). The background model is rendered on the GPU and the keying is done by thresholding the depth values in a GPU shader. The segmented dynamic person is also represented as an octree.

While the environment model is computed offline once, the dynamic person model is updated in each frame as Space-Time-Video to reflect the object motion and space occupancy. Figure 9 shows the octree volume representation for one of the frames, including the person's model. Animation and collision detection of computer generated elements allow for dynamic interaction of the computer generated elements with the real scene. As an example, colored balls are dropped into the scene while the real person is walking around. The virtual balls are colliding and bounce off the real scene objects. Since the camera observes the frontal object surface only, there might be some missing collisions at the person's back side, but that does not really harm the visual effect. Figure 9 shows two frames of the resulting animation sequence. The balls are reflected and even stirred up by the legs of the walking person.

6 Conclusion

In this work we focused on the discussion and use of data structures for the representation of ToF-camera data. We defined several requirements that a data structure has to provide and we concluded that the octree offers best possibilities. We evaluated several scenarios such as environment reconstruction, model building, Space-Time-Video and interactive collision detection. For environment reconstruction, octrees can compete with 2.5D panoramic image, although image blending computes smoother surfaces. For the other scenarios, a full 3D structure like the octree is necessary.

References

1. Oggier, T., Lehmann, M., Kaufmann, R., Schweizer, M., Richter, M., Metzler, P., Lang, G., Lustenberger, F., Blanc, N.: An all-solid-state optical range camera for 3d real-time imaging with sub-centimeter depth resolution. In: Proceedings of SPIE, vol. SPIE-5249, pp. 534–545 (2003)
2. Xu, Z., Schwarte, R., Heinol, H., Buxbaum, B., Ringbeck., T.: Smart pixel - photonic mixer device (pmd). In: M2VIP - International Conference on Mechatronics and Machine Vision in Practice, pp. 259–264 (1998)
3. Hartley, R.I., Zisserman, A.: Multiple View Geometry in Computer Vision. Cambridge University Press, Cambridge (2000)
4. Kolb, A., Barth, E., Koch, R., Larsen, R.: Time-of-Flight Sensors in Computer Graphics. In: Proceedings of Eurographics (State-of-the-Art Report), Munich, Germany (March 2009)
5. Lindner, M., Kolb, A.: Lateral and depth calibration of PMD-distance sensors. In: Bebis, G., Boyle, R., Parvin, B., Koracin, D., Remagnino, P., Nefian, A., Meenakshisundaram, G., Pascucci, V., Zara, J., Molineros, J., Theisel, H., Malzbender, T. (eds.) ISVC 2006. LNCS, vol. 4292, pp. 524–533. Springer, Heidelberg (2006)
6. Lindner, M., Kolb, A.: Calibration of the intensity-related distance error of the PMD ToF-camera. In: SPIE, Intelligent Robots and Computer Vision, vol. 6764 (2007), doi:10.1117/12.752808
7. Schiller, I., Beder, C., Koch, R.: Calibration of a pmd camera using a planar calibration object together with a multi-camera setup. In: The International Archives of the Photogrammetry, Remote Sensing and Spatial Information Sciences, Beijing, China, vol. XXXVII, Part B3a, pp. 297–302 XXI (2008); ISPRS Congress
8. Lindner, M., Schiller, I., Kolb, A., Koch, R.: Time-of-flight sensor calibration for accurate range sensing. In: Computer Vision and Image Understanding CVIU, Special Issue on Time-of-Flight Camera based Computer Vision (2009) (accepted for publication)
9. Bartczak, B., Schiller, I., Beder, C., Koch, R.: Integration of a time-of-flight camera into a mixed reality system for handling dynamic scenes, moving viewpoints and occlusions in real-time. In: Proceedings of the 3DPVT Workshop, Atlanta, GA, USA (June 2008)

10. Huhle, B., Jenke, P., Straßber, W.: On-the-fly scene acquisition with a handy multisensor-system. International Journal of Intelligent Systems Technologies and Applications 5(3-4), 255–263 (2008)
11. Prusak, A., Melnychuk, O., Schiller, I., Roth, H., Koch, R.: Pose estimation and map building with a pmd-camera for robot navigation. International Journal of Intelligent Systems Technologies and Applications 5(3-4), 355–364 (2008)
12. Shade, J., Gortler, S., He, L.w., Szeliski, R.: Layered depth images. In: SIGGRAPH 1998: Proceedings of the 25th annual conference on Computer graphics and interactive techniques, pp. 231–242. ACM Press, New York (1998)
13. Curless, B., Levoy, M.: A volumetric method for building complex models from range images. In: SIGGRAPH 1996: Proceedings of the 23rd annual conference on Computer graphics and interactive techniques, pp. 303–312. ACM, New York (1996)
14. Carlbom, I., Chakravarty, I., Vanderschel, D.: A hierarchical data structure for representing the spatial decomposition of 3-d objects. IEEE Computer Graphics and Applications 5(4), 24–31 (1985)
15. Szeliski, R.: Rapid octree construction from image sequences. CVGIP: Image Underst. 58(1), 23–32 (1993)
16. Meagher, D.: Geometric modeling using octree encoding. Computer Graphics and Image Processing 19(2), 129–147 (1982)
17. Rusinkiewicz, S., Levoy, M.: Qsplat: a multiresolution point rendering system for large meshes. In: SIGGRAPH 2000: Proceedings of the 27th annual conference on Computer graphics and interactive techniques, pp. 343–352. ACM Press/Addison-Wesley Publishing Co. (2000)
18. Samet, H.: Implementing ray tracing with octrees and neighbor finding. Computers And Graphics 13, 445–460 (1989)
19. Shi, J., Tomasi, C.: Good features to track. In: IEEE Conference on Computer Vision and Pattern Recognition, Seattle, June 1994, pp. 593–600 (1994)
20. Evers-Senne, J.F., Koch, R.: Image-based rendering of complex scenes from a multi-camera rig. In: IEEE Proceedings Vision Image and Signal Processing. vol. 152(4) (August 2005)
21. McMillan, L., Bishop, G.: Plenoptic modeling: an image-based rendering system. In: SIGGRAPH 1995: Proceedings of the 22nd annual conference on Computer graphics and interactive techniques, pp. 39–46. ACM, New York (1995)
22. Gorelick, L., Blank, M., Shechtman, E., Irani, M., Basri, R.: Actions as space-time shapes. In: ICCV, pp. 1395–1402. IEEE, New York (2005)

Fusing Time-of-Flight Depth and Color for Real-Time Segmentation and Tracking

Amit Bleiweiss and Michael Werman

School of Computer Science
The Hebrew University of Jerusalem
Jerusalem 91904, Israel
{amitbl,werman}@cs.huji.ac.il

Abstract. We present an improved framework for real-time segmentation and tracking by fusing depth and RGB color data. We are able to solve common problems seen in tracking and segmentation of RGB images, such as occlusions, fast motion, and objects of similar color. Our proposed real-time mean shift based algorithm outperforms the current state of the art and is significantly better in difficult scenarios.

1 Introduction

Segmentation and tracking are important basic building blocks of many machine vision applications. Both problems have been the subject of research for many years. In order to understand an image, it is often crucial to be able to segment it into separate objects before proceeding with further analysis. The mean shift algorithm is a popular scheme widely used today to solve these two problems. Mean shift has a clear advantage over other approaches, in that the number of clusters does not need to be known beforehand (as in GMMs and K-Means), and it does not impose any specific shape on the clusters.

In [5], a robust framework for segmentation by combining edge and color data was presented and [4] applied mean shift to tracking of objects in RGB color space. However, both of these approaches suffer from the usual problems associated with the analysis color images. Color-based segmentation does not work well when the background has a similar color to objects in the scene. Tracking typically relies on a color based histogram and thus performs poorly when lighting changes, or when close objects share the same color. Improvements have been made by adding edge information, a Kalman prediction step [6] or modifying the kernel applied to the image samples [2], but they still have the limitation that these measures rely on the same color data. Therefore, it is desireable to add additional information to the problem at hand [12] in order to get an improvement, and thus fusing time-of-flight depth data with color can lead to much better results.

Several other papers deal with fusion of depth with other data derived from a camera. [16] fuses high-resolution color images with time-of-flight data. [10]

R. Koch and A. Kolb (Eds.): Dyn3D 2009, LNCS 5742, pp. 58–69, 2009.

and [11] use low-resolution time of-flight data to fix discontinuities in the high-resolution depth data acquired from a stereo camera setup. [7] fused time-of-flight and color data to segment the background in a video sequence, though the performance was about 10 frames per second. [8] fused laser range and color data to train a robot vision system. In [18], local appearance features extracted from color data are fused with stereo depth data in tracking of objects. Haar-like features detected in the color data are used to improve noise and low-confidence regions inherent in depth data produced from stereo cameras. However, the algorithm runs at 15Hz on a powerful PC due to the amount of computation necessary for extracting the appearance models. In [17], two separate particle filter trackers, one using color and the other using time-of-flight data, are described and compared on a variety of video sequences. Again, the approach for both trackers is not suitable for real time as it involves heavy processing. The paper concludes that each performs better in different environments and that combining the two would be beneficial.

In fact, it is a very natural step to use depth data to compensate for noise and problematic artifacts present in other data. In the case of standard digital camera output, the time-of-flight depth data naturally compensates for the disadvantages and weaknesses of RGB data. For example, the depth image automatically segments background from foreground objects, a very difficult task in color images. It is also mostly unaffected by light sources, whereas RGB images recorded with a standard digital camera change color when lighting changes. On the other hand, depth data tends to be noisy and contain artifacts which need to be handled.

In this paper we show that the natural fusion of color and depth data can yield much improved results in both segmentation and tracking. In section 2 we give an overview of the data used. In section 3 we investigate sensor fusion in applying mean shift segmentation to a single image. In section 4 we extend this idea to tracking objects in real time using mean shift. In section 5 we state our conclusions and future work.

2   Data Stream

For our data, we used 3DV Systems' ZSense camera [9], which simultaneously captures both RGB and depth data using the time-of-flight principle. This gives us a 4-channel image with 8 bits per channel. Although the data is recorded using 2 separate lenses, the RGB image is already warped to match the depth image. Thus, there is no need for calibration or matching of the data as in [16]. The ZSense camera has several options for resolution and frequency, and we used 320x240 resolution captured at 30fps. The camera has a range of 0.5-2.5m, a 60° field of view, with a depth accuracy of a few cm., and RGB data comparable to a standard webcam.

3   Mean Shift Segmentation

Although the segmentation problem in image processing is not well defined, our goal is to cluster the pixels of an image into separate objects in a robust manner, regardless of the fact that separate objects may share identical colors. Color-based segmentation does not work well in cases of occlusion, or when objects in the scene share a similar color with their background. We need to introduce additional data to the problem, and thus the fusion of time-of-flight depth data fits naturally in this scheme. However, adding the depth data blindly as an additional dimension does not necessarily improve the results. Depth data is noisy and may contain artifacts which can lead to worse results. Thus we must fuse the data wisely, by deciding when to use depth, and how much weight it should be given. Usually depth data should carry more weight, as it is lighting independent, and thus objects in the scene will not share the same color as the background, but we have to be careful due to the depth sensor's noise.

3.1   Proposed Algorithm

In [5] the mean shift procedure is applied iteratively to every point in a 5D feature space. The feature space consists of the color (converted to L*u*v* color space for a better distance metric) and the 2D lattice coordinates. At each iteration, the window is moved by a 5D mean shift vector, until convergence, (when shifts are smaller than a given threshold). In order to modify this algorithm we use depth data as an additional dimension to the feature space, yielding clusters which are similar in color as well as in 3D Euclidean space. Thus we extend the above approach by computing a 6D mean shift vector per iteration. The algorithm does the following:

1. Convert the color data to L*u*v*.
2. Estimate the noise in the depth data. We use a simple scheme similar to [13], by applying a smooth function to the depth image D, and then approximating the residual of the smooth image S and the original image D as noise.

$$W = |S - D| \tag{1}$$

   This approach yields the over-estimated noise matrix W, but works well in practice. We use the bilateral filter [15] as a smoother to preserve edges while removing many unwanted artifacts present in the depth data. (see Fig. 1)
3. When computing the 6D mean shift vector, we scale the original weights derived in [5] by W, and add an additional scale factor σ so as to give more overall weight to the depth data when it is not noisy.

$$w_6(x) = \frac{1}{W(x) + 1} w_6(x)\sigma \tag{2}$$

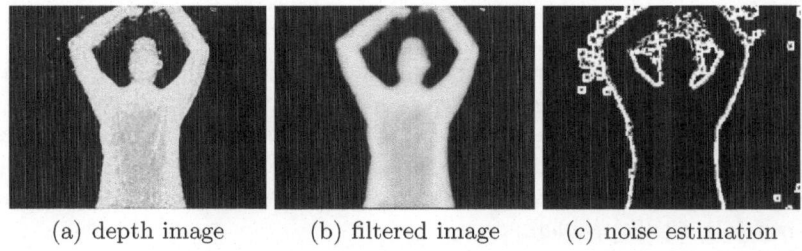

(a) depth image   (b) filtered image   (c) noise estimation

Fig. 1. Noise estimation of depth using bilateral filter

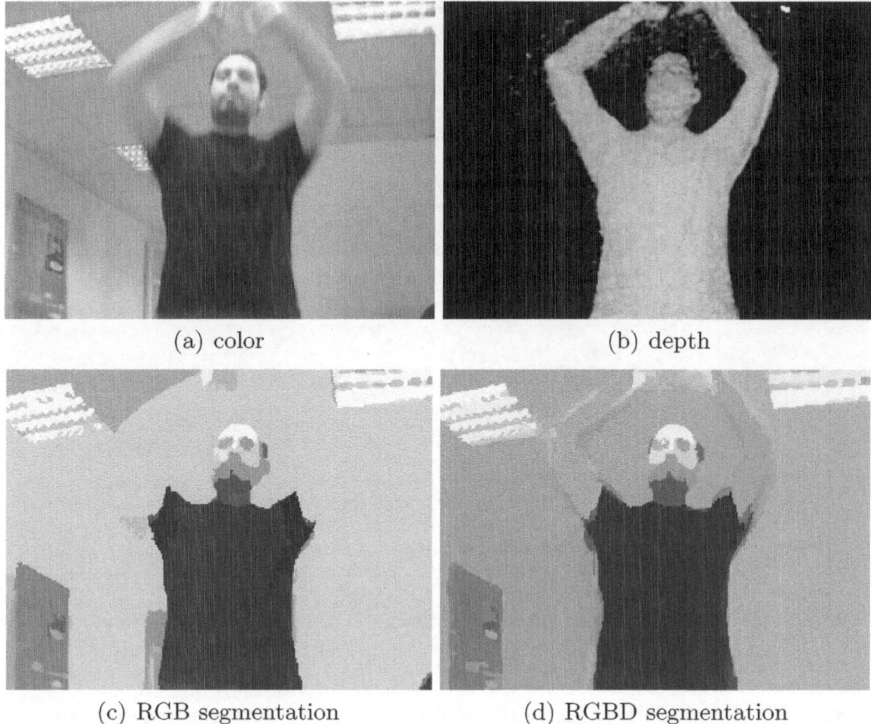

(a) color   (b) depth

(c) RGB segmentation   (d) RGBD segmentation

Fig. 2. Segmentation of image from a fast motion sequence. Note that the arms in (c) are clustered with the background due to noisy color data. Results are significantly improved in (d).

where w_i is the weight applied to each component of the mean shift vector (XYRGBD), and w_6 is the weight applied to the depth component. In practice, σ values in the range of 2-5 yielded the best results. In this manner, mean shift will rely more on the color data in regions where the depth data is noisy.

3.2   Results

This scheme performs much better than the pure color-based scheme. In a sequence recorded with fast motion (see Fig. 2), the color data is "smeared" causing the person's arms to be incorrectly clustered with the ceiling. Adding the weighted depth data solves the problem. In an occlusion sequence (see Fig. 3), parts of the person are incorrectly clustered with the background, whereas adding the depth solves this problem.

(a) color	(b) depth
(c) RGB segmentation	(d) RGBD segmentation

Fig. 3. Segmentation of image with partial occlusion. Note the person's right arm and left shoulder in (c) are incorrectly clustered with the background. Results are significantly improved in (d).

4   Mean Shift Tracking

Comaniciu et. al. [4] extend the idea presented in [5] to robust tracking of objects in time. The mean shift procedure is now applied locally to a window until it converges on the most probable target. The main idea is to compute an initial histogram in a small window for the object we wish to track, and then use mean shift to find where to move the window so that the distance between the histograms is minimized. While this scheme is robust, we note that it suffers from the same disadvantages of other color-based approaches. Naturally, it cannot track well when objects share a similar color with the background. For example, the tracker does not perform well when a wall in the background is similar to the person's skin color. On the other hand, relying purely on the depth data has its own share of problems. For example, when hands are close to the body, there is no variance in the depth window, and thus we have no way to locally track the hands. It is clearly desirable to fuse the depth and RGB data in order to achieve optimal results. In the following sections we describe the proposed tracking algorithm.

4.1 Histograms

Our tracking uses a local mean shift procedure, and we have the freedom to use different data for each frame of the sequence. Thus we are able to use different histograms for each new frame depending on several measures in the local window. The main challenge is to decide when and how to fuse the depth and color information. This is important for two reasons:

1. In most cases we wish to use both color and depth information. However, in some cases we may want to drop one of these channels as it can actually degrade the results. For example, if a white ball is thrown in front of a white wall, we wish to only use the depth data. On the other hand, when we are tracking hands which are very close to a person's body, it is desirable to drop the depth data. Thus we need a good set of rules to decide what to fuse.
2. Since we are working with a 32 bit image using a bin for each possible color+depth will yield a histogram of 4 billion bins. Adding additional local descriptors for robustness [14,1] will make the number of bins even larger. The number of possible colors at each local iteration is much smaller, so using the above logic we would get a very sparse histogram. Clearly, we must quantize each channel, but also drop undesired data to save memory.

We initially ran the algorithm naively with a RGBD histogram of 16x16x16x16 bins, and made the following observations:

1. For most sequences, we get better results immediately.
2. We get worse performance in cases where depth data in the target window is noisy. We can use our weight map from the previous section in order to decide if the depth pixels in the target window are noisy, and then rely only on the color data.
3. When there is not enough information to track well, the number of mean shift iterations increases significantly. We noticed this both in fast motion where RGB data was blurry, as well as in cases where both depth and RGB data in the window neighborhood had low variance. In this case, we can double the number of bins and use additional robust image descriptors as well.

Using the above observations, we are now able to apply a robust algorithm which fuses the depth and color data in an optimal manner.

4.2 Proposed Algorithm

We use a framework similar to [4], but modify the histogram. Given a distribution q of the target model and an initial location y_0

1. Convert the color data to L*u*v*
2. Apply a bilateral filter to the depth data and compute weight matrix W (eq. 1)
3. The histogram is computed using color and depth data, quantized so as to fit in less bins. Each channel is divided by 16, yielding a total of 16^4 possible bins instead 256^4. If the number of depth pixels in the weight map is above a threshold, we only use a 16^3 bin histogram and drop the depth data. In this way we ignore noise in the depth data.
4. Compute p, the distribution for the target model at y_0 using

$$\hat{p}(y) = C_h \Sigma_{i=1}^{n_h} k(\|\frac{y - x_i}{h}^2\|)\delta[b(x_i) - u] \tag{3}$$

where k is the Epanechnikov kernel, h is the window radius, $\{x_i\}_{i=1...n_h}$ are the pixel locations of the target object, δ is the Kronecker delta function, and b is a function associating a histogram bin with a given pixel. p and q are density functions for the color/depth feature of the target areas, and thus we want to find the target position y at which the density functions are most similar.

5. Evaluate the Bhattacharyya distance [4] between the 2 distributions

$$\rho[\hat{p}(\hat{y}_0), \hat{q}] = \Sigma_{u=1}^m \sqrt{(\hat{p}_u(y_0)\hat{q}_u)} \tag{4}$$

This distance is a measure we want to minimize in order to pick the best target candidate

6. Derive the weights for each pixel

$$w_i = \Sigma_{u=1}^m \delta[b(x_i) - u]\sqrt{\frac{\hat{q}_u}{\hat{p}_u(y_0)}} \tag{5}$$

7. Compute the mean shift vector (target displacement)

$$\hat{y}_1 = \frac{\Sigma_{i=1}^{n_h} x_i w_i g(\|\frac{\hat{y}_0 - x_i}{h}^2\|)}{\Sigma_{i=1}^{n_h} w_i g(\|\frac{\hat{y}_0 - x_i}{h}^2\|)} \tag{6}$$

where $g(x) = -k'(x)$, and use it to find the new estimated y position.

8. Repeat steps 4-7 until convergence (displacement < threshold).
   (a) In most cases, the fusion of RGB and depth data is sufficient, and we get and average of 5 iterations for convergence.
   (b) In the case of low variance in the window region, we typically get a much larger number of iterations, causing the tracker to sometimes lose the object. In order to offset this effect, we keep track of a slightly larger window of radius l surrounding the original tracking window. If the variance of

the RGB data is low within this window, we run a SURF detector [1] in the local window and recompute the histogram function b. The number of bins is doubled in this case, as each pixel needs to indicate whether it contains a SURF descriptor or not. In practice, runtime performance is not largely impacted as this case does not happen often.

4.3   Handling Extreme Cases

We would like our framework to also handle extreme cases, such as turning off all lights during a sequence, or having a person walk out of the camera's depth sensor range. In the previous section, we described how to ignore depth data in regions containing noisy pixels. We wish to extend this idea in order to handle other cases where either RGB or depth data is to be ignored in the histogram. Again, we keep track of a larger window of radius l surrounding the actual tracking window. At each iteration of the mean shift procedure, we check this window and compute the sum of the pixel values for each data type respectively:

$$S_{rgb} = \Sigma_{i=1}^{n_l}(R(x_i) + G(x_i) + B(x_i)) \qquad (7)$$

$$S_{depth} = \Sigma_{i=1}^{n_l} D(x_i) \qquad (8)$$

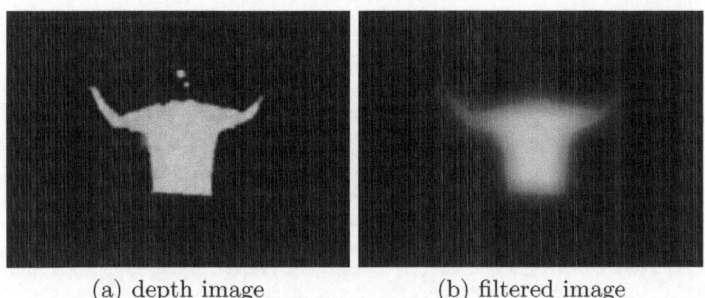

(a) depth image   (b) filtered image

Fig. 4. Out of range depth data

Table 1. Performance Statistics (for sequences of 1000 frames)

Sequence Description	# of frames using RGB+D	# of frames using RGB only	# of frames using additional SURF data
slow motion, no occlusion	972	28	0
occlusion with moving hands	914	86	7
occlusion between hand and head	926	74	23
rolled-up sleeves, occlusions	989	11	247
person standing close to wall	752	248	0

1. In the case of people walking out of the camera's range, a vast majority of the depth pixels are zero. Thus, when S_{depth} is close to 0, only RGB data is used in the histogram.
2. In the case of no lights, all pixels in the RGB image are zero. Thus, when S_{rgb} is close to 0, only depth data is used in the histogram.

This simple heuristic works better on the bilateral filtered depth image, as we get some noisy pixels when people are close to the depth sensor's far limit. The filter removes this noise, and works better in practice (see Fig. 4)

4.4   Results

We tested our algorithm on a set of challenging sequences with both occlusions, proximity to walls, and fast motion. In all cases, our algorithm outperformed the color based framework. We show results for a subset of our sequences, focusing on hand tracking in different environments and with complex motions (See table 1) Fig.5 shows successful tracking of hands completely occluding each other. Fig. 6 shows an example in which hands were close to the body and sleeves were rolled up, causing problems for the color tracker. Additional SURF descriptors were sometimes added to the histogram in order to successfully track the hands. Fig. 7 shows the number of iterations per frame for a typical sequence. Note that most frames take a small number of iterations, and additional robust descriptors were added only for the extreme peaks, improving the tracking results.

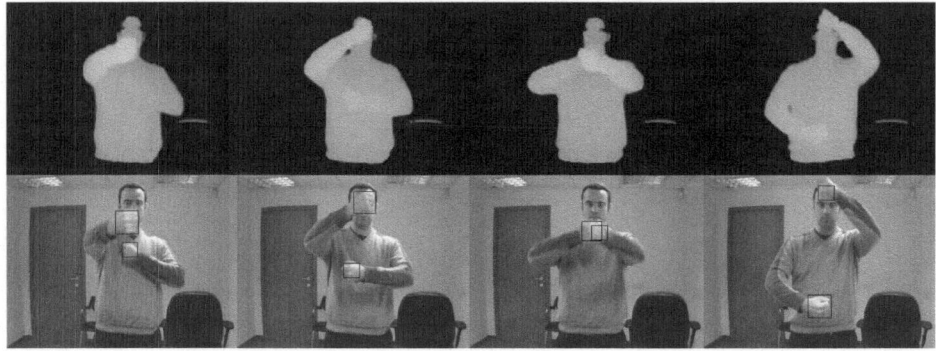

Fig. 5. Tracking a sequence with occlusions (a) initial tracking window is intentionally placed close to the body on left hand to make the tracking harder as variance is low in the depth image in that window (b) hand is occluding the face, both having a similar color histogram. Tracking works well with additional depth data, but stays on face when relying on color only. (c) hands completely occlude each other but continue to track well afterwards when depth information is added to the histograms. (d) tracking continues successfully even when close to body, a case in which tracking based on depth only will fail.

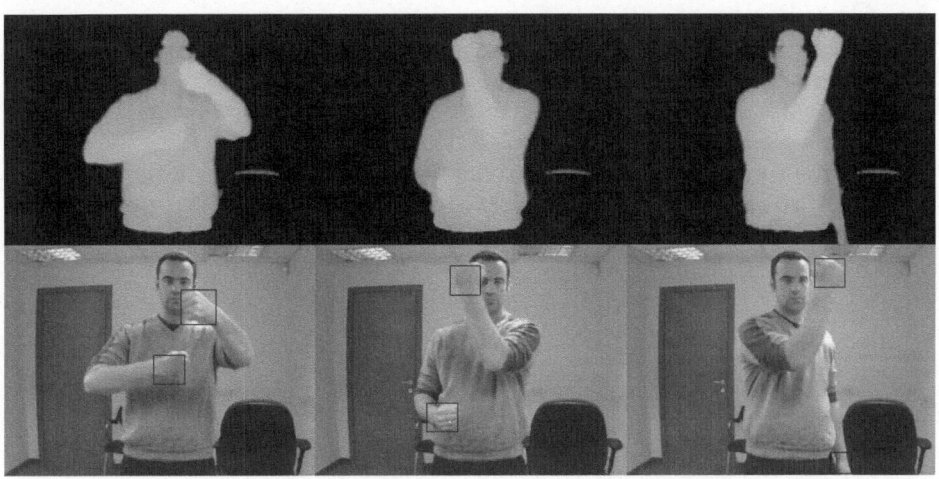

Fig. 6. Tracking a sequence with rolled-up sleeves, yielding a large region with similar histogram to that of the target window. The images above show cases in which both color and depth have a large region with similar depth and color data. Using a naive RGBD tracker caused instability with the window, moving all over the arm region. In this case, mean shift results in many iterations. Adding the additional robust SURF descriptors results in successful tracking throughout the sequence.

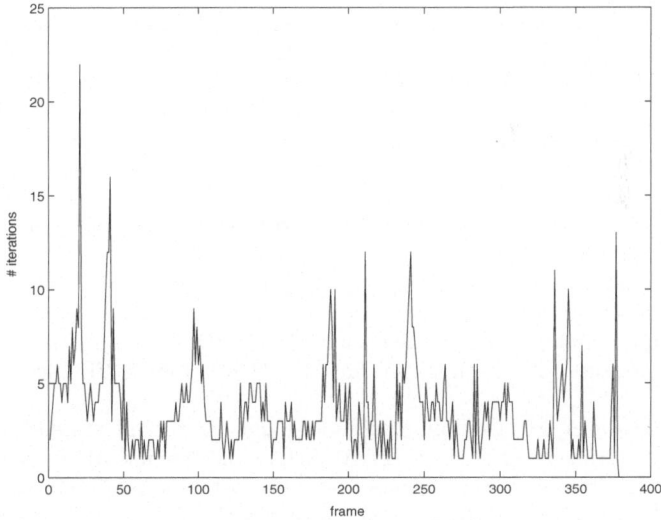

Fig. 7. Number of mean shift iterations per frame for a sample sequence

5   Conclusions

We have shown how fusion of depth data to color images results in significant improvements in segmentation and tracking of challenging sequences. The proposed algorithm runs at 45 fps on a single-core 2.4GHz PC, and the number of iterations is similar to that in the original mean shift implementation.

The algorithm can be further improved by using the robust descriptors generated by SURF [1] rather than just the detected points, and incorporating them within the Bhattacharyya distance framework.

Acknowledgements

The authors would like to thank Omek Interactive for providing the sequence data.

References

1. Bay, H., Ess, A., Tuytelaars, T., Van Gool, L.: SURF: Speeded Up Robust Features. Computer Vision and Image Understanding 110(3), 346–359 (2008)
2. Chen, Z., Husz, Z., Wallace, I., Wallace, A.: Video Object Tracking Based on a Chamfer Distance Transform. In: Proceedings of IEEE International Conference on Image Processing, pp. 357–360 (2007)
3. Collins, T.: Mean-Shift Blob Tracking through Scale Space. In: CVPR, vol. 2, pp. 234–240 (2003)
4. Comaniciu, D., Ramesh, V., Meer, P.: Real-Time Tracking of Non-Rigid Objects using Mean Shift. In: CVPR, vol. 2, pp. 142–149 (2000)
5. Comaniciu, D., Meer, P.: Mean Shift: A Robust Approach towards Feature Space Analysis. IEEE Trans. Pattern Analysis Machine Intell. 24(5), 603–619 (2002)
6. Comaniciu, D., Ramesh, V.: Mean Shift and Optimal Prediction for Efficient Object Tracking. In: International Conference on Image Processing, pp. 70–73 (2000)
7. Crabb, R., Tracey, C., Puranik, A., Davis, J.: Real-time Foreground Segmentation via Range and Color Imaging. In: CVPR Workshop on time-of-flight Camera Based Computer Vision (2008)
8. Gould, S., Baumstarck, P., Quigley, M., Ng, A., Koller, D.: Integrating Visual and Range Data for Robotic Object Detection. In: M2SFA2 2008: Workshop on Multi-camera and Multimodal Sensor Fusion (2008)
9. Gvili, R., Kaplan, A., Ofek, E., Yahav, G.: Depth Key. SPIE Electronic Imaging (2006)
10. Hahne, U., Alexa, M.: Combining time-of-flight Depth and Stereo Images without Accurate Extrinsic Calibration. In: Workshop on Dynamic 3D Imaging (2007)
11. Kuhnert, K., Stommel, M.: Fusion of stereo-camera and pmd-camera data for real-time suited precise 3d environment reconstruction. In: IEEE/RSJ International Conference on Intelligent Robots and Systems, pp. 4780–4785 (2006)
12. Leichter, I., LindenBaum, M., Rivlin, E.: A General Framework for Combining Visual Trackers - The Black Boxes Approach. International Journal of Computer Vision 67(3), 343–363 (2006)
13. Liu, C., Freeman, W., Szeliski, R., Bing Kang, S.: Noise Estimation from a Single Image. In: CVPR, vol. 1, pp. 901–908 (2006)

14. Lowe, D.G.: Object recognition from local scale-invariant features. In: Proceedings of the International Conference on Computer Vision, pp. 1150–1157 (1999)
15. Paris, S., Kornprobst, P., Tumblin, J., Durand, F.: A Gentle Introduction to Bilateral Filtering and its Applications. ACM Siggraph Course Notes (2008)
16. Reulke, R.: Combination of Distance Data with High Resolution Images. In: IEVM Proceedings (2006)
17. Sabeti, L., Parvizi, E., Wu, J.: Visual Tracking Using Color Cameras and time-of-flight Range Imaging Sensors. Journal of Multimedia 3(2), 28–36 (2008)
18. Tang, F., Harville, M., Tao, H., Robinson, I.N.: Fusion of Local Appearance with Stereo Depth for Object Tracking. In: CVPR, pp. 142–149 (2008)

Depth Imaging by Combining Time-of-Flight and On-Demand Stereo

Uwe Hahne and Marc Alexa

TU Berlin, Germany
hahne@cs.tu-berlin.de, marc.alexa@tu-berlin.de
http://www.cg.tu-berlin.de

Abstract. In this paper we present a framework for computing depth images at interactive rates. Our approach is based on combining time-of-flight (TOF) range data with stereo vision. We use a per-frame confidence map extracted from the TOF sensor data in two ways for improving the disparity estimation in the stereo part: first, together with the TOF range data for initializing and constraining the disparity range; and, second, together with the color image information for segmenting the data into depth continuous areas, enabling the use of adaptive windows for the disparity search. The resulting depth images are more accurate than from either of the sensors. In an example application we use the depth map to initialize the z-buffer so that virtual objects can be occluded by real objects in an augmented reality scenario.

1 Introduction

Real time depth imaging is a building block in many interactive vision systems and, in particular, is necessary for enabling realistic occlusions in augmented reality (AR). Despite the improved speed of general purposes computing as well as development of new types of sensors, providing depth images in *real time* continues to be a challenging problem. For the purposes of enhancing AR with convincing occlusions current approaches are limited to either reducing quality in the depth maps [SNV02] or realizing occlusions by compositing [VH08]. We are demonstrating a first step toward an affordable and lightweight solution by fusing information from a low cost but also low resolution time-of-flight range sensor with standard correlation-based stereo.

Quite generally, we can distinguish active and passive approaches to real time depth imaging. Active optical techniques involve relighting the scene and usually require an expensive and heavy setup. Recently, sensors based on the time-of-flight principle for sensing depth have become affordable and fast with the introduction of photonic mixer devices (PMD) [MKF+05, XSH+05]: the reflection of modulated IR light is collected in a CMOS matrix. Comparing the signal to the source modulation yields the phase, which is a linear function of distance to the reflecting surface. PMD depth imaging works at interactive rates, but suffers from comparably low spatial resolution of the sensor and noise in the depth values, especially for surfaces with low reflectance.

Passive techniques, at least when several frames per second are required, are based on multiple views of scene captured with two or more cameras. We have decided to

R. Koch and A. Kolb (Eds.): Dyn3D 2009, LNCS 5742, pp. 70–83, 2009.

use a single binocular stereo camera, as more not only make the setup more complicated, but also require processing more images. From the large number of stereo vision
approaches [SS02] only local correlation based methods are fast enough for real time
application [FHM+93, Hir01]. This limits the quality of the resulting depth maps, most
obviously in large featureless areas but also at depth discontinuities, where the correlation window might compare different objects because of occlusion.

We combine the camera systems (time-of-flight and stereo) and fuse the data so that
limitations of each of the individual sensors are compensated. Our goal is enhancing the
high resolution color image from one of the stereo camera oculars with depth information, gathered from the PMD camera as well as from disparity estimation. The mapping
of the PMD depth image into a color image acquired with another camera (resp. texturing the depth data with the color image) has been analyzed by Reulke [Reu06] and
Lindner et al. [LKH07,LLK07]. Our setup combines the PMD camera with a binocular
stereo camera, similar to [BBK07, GAL07, HA07, NMCR08, KS06, ZWYD08].

For better explaining our choice of algorithm, we need to briefly touch on the setup,
calibration, and properties of the cameras (section 2). The physical properties of the
PMD camera give rise to the preprocessing of its data, most importantly the estimation
of confidence values for each depth value (section 3). Kuhnert and Stommel [KS06],
as well as Netramai et al. [NMCR08] use a similar confidence map to choose either
the depth value acquired with the PMD camera or depth from stereo – we exploit this
depth/confidence map for initializing and steering a local correlation based stereo algorithm (section 4), in particular by choosing adaptive windows for the correlation based
on the information in both the color images and the range image.

In an earlier approach [HA07], we combined the TOF data from the PMD camera
with high resolution images from two photo cameras, using graph cuts to find a globally
optimal solution for a depth map of a single perspective. The use of graph cuts leads to
computation times that are insufficient for real-time video processing. Similarly, Guo-
mundsson et al. [GAL07], Zhu et al. [ZWYD08], and Beder et al. [BBK07] generate
depth images by fusing TOF and stereo data. Their approaches appear to be much faster
than using graph cuts, however, they target single images and provide no information on
the computation times. The choices of stereo algorithm, however, indicate that they are
not amenable to real-time processing in their current form. We explicitly start from the
restrictive setting of real-time applications, which severely restricts the choice of stereo
algorithm, mostly to local correlation with fixed windows. We use the TOF information
particularly to adapt the windows, as fixed windows fail at depth discontinuities. We
believe our approach yields depth images at interactive rates

- that are are more reliable than the information from the PMD camera without com-
  promising the interactive frame rate and
- that are more accurate around depth discontinuities than real time stereo vision
  approaches based on fixed window correlation.

We demonstrate our use of the system in an augmented reality (AR) scenario for
computing accurate occlusions between virtual and real objects.

2   Setup

We mount a compact time-of-flight camera PMDTec type [vision]19k, capturing a depth range of about 7.5 meters at a resolution of 160x120 pixels, together with PointGrey Bumblebee2 stereo camera capturing color video at a resolution of 640x480, on an aluminum rack (see Figure 1 for an image of the cameras). Both cameras are aligned to parallel viewing directions.

2.1   Photonic Mixer Device Depth Camera

Photonic Mixer Devices (PMD) are semiconductor sensors that can be used to measure distances per pixel based on the time-of-flight principle [XSH$^+$05, MKF$^+$05]. The camera system includes a light source based on infrared LEDs, illuminating the scene with a continuously modulated signal at $f = 20$MHz. The sensor detects the phase shift between the source and received signal by sampling four values per period. The phase shift is in principle independent of the amplitudes of the signals, and since modulation frequency f and speed of light c are constants, it relates linearly to the distance of the reflecting object.

Measuring phase shifts has several inherent limitations: first, phase shifts have symmetries along the signal and are unique only in an interval of π, in this case $c/2f =$ 7.5m. More subtly, measuring phase shifts assumes perfectly sinusoidal signals, yet this is not the case. This leads to a "wiggling" error in the depth measurement (see Rapp et al. [Rap07]). Most importantly, phase shifts can only be measured accurately if the sensor receives the right amount of light. If objects are too dark sensor noise dominates the signal, while too much light leads to saturation and makes modulation undetectable. As a consequence, measurements depend on object reflectance and distance to the light source and camera. The camera allows controlling the integration time (similar to

Fig. 1. All three cameras mounted on an aluminum bar

exposure time in standard cameras), however, just as in photography it is often impossible to keep all objects captured in the scene within the dynamic range of the sensor.

The camera provides the raw signal samples as well as consolidated amplitude and depth values computed from the four samples. Interestingly, the consolidated amplitude value is accurate only as long as the sensor is not being saturated. It is possible to detect saturation based on the raw signal samples (see Rapp et al. [Rap07]).

The resulting depth images captured with PMD technology are noisy and contain wrong depth values because of the phase ambiguity and objects reflecting not enough light; whereas we have eliminated the case of reflecting too much light by adjusting the integration time.

2.2 Calibration

Sensor fusion requires registration and accurate calibration. For both the intrinsic and extrinsic calibration we use the calibration algorithm of Zhang [Zha99, Zha00]. The PMD camera, however, yields intensity images that are too noisy for direct application and they are preprocessed following the ideas of Reulke [Reu06] as well as Lindner and Kolb [LK06]. A relevant practical problem for the extrinsic calibration is the misalignment of optical center and zero depth plane of the PMD camera. Interestingly, Gudmundsson et al. [GAL07] perform a stereo calibration between all pairs of cameras. We have found this to be cumbersome, because of the combination of noisy amplitude images and the mismatch between optical center and depth image for the PMD camera. We rather consider only the extrinsic calibration between the depth image from PMD camera and the systems of the color cameras.

The stereo camera color images are rectified to reduce the correspondence problem to a single line. Our calibration is accurate enough so that the depth difference between stereo system and the PMD camera is within the accuracy of the PMD camera.

3 Preprocessing and Confidence Estimation

As explained in the last section, the quality of the depth values captured by the PMD camera depends strongly on the surface of the objects in the scene. Dark and glossy surfaces lead to artifacts as the modulated IR light is not reflected as expected. Especially when using the depth values for determining occlusions in AR applications, these artifacts become clearly visible. We process the data prior to using it with the stereo system, trying to improve the data by simply filtering and assigning confidence values to each depth value. Very low confidence depth values are replaced by interpolated values with higher confidence.

3.1 Filtering

Reducing the noise or removing outliers is one obvious part of the pre-process. Isolated outliers can be removed at minimal cost using median filtering. Through experimentation we have found that a small kernel of 3×3 pixel is sufficient. This appears to be due to the outliers being mostly isolated pixels. Larger kernels would lead to longer processing times without showing a significant improvement.

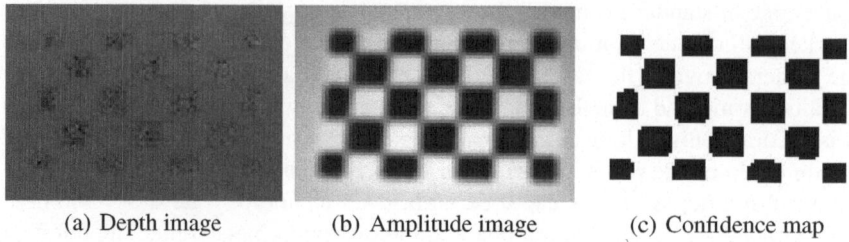

(a) Depth image   (b) Amplitude image   (c) Confidence map

Fig. 2. A checkerboard is difficult to reconstruct using PMD range sensing, because of the insufficient amount of light reflected by the black areas. The acquired depth image (a) clearly holds wrong depth values in these black areas. The amplitude image (b) can be used to compute a confidence map (c), which is thresholded to classify depth values as valid (white) or invalid (black)

3.2 Confidence Estimation

As explained in the last section, in our setup, the dominant cause for systematically wrong depth estimation are objects that have bad reflection properties due to their material and color. However, this information is available in form of an amplitude image of the scene.

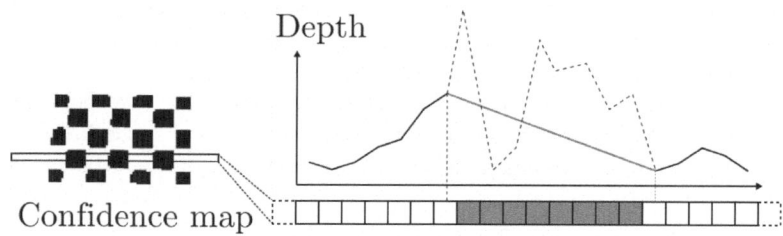

Fig. 3. Scanline interpolation of the PMD data using the confidence map. The red line is the interpolated depth, the dashed line is the original unreliable PMD depth.

The first step in turning the amplitude image into a confidence map is applying a 3×3 median filter, similar to the process for the depth image. The resulting image is thresholded, yielding a binary confidence map classifying depth values as either valid or invalid. Figure 2 highlights the problems resulting from low reflectivity at the example of a checkerboard and shows the resulting classification of valid and invalid depth values (while the 3D reconstruction from the wrong depth values can be seen in Fig. 4(a)).

3.3 Interpolation

In all regions marked as confident by the binary map, we will use depth values for initializing disparities. In most cases the depth values in insufficiently reflecting areas are far from the truth and it is better to assume continuity in depth.

(a) Original image   (b) With interpolation

Fig. 4. 3D reconstruction of the scene from the depth image (and using the intrinsic camera geometry). The left image shows the reconstruction from the data in Figure 2(a) and the right image uses linearly interpolated depth values for elements with low confidence.

As in our case the interpolated depth values will later be corrected using the stereo information, we opt for a simple approach that is as fast as possible: the depth image is scanned across horizontal lines. When an invalid segment is encountered it is replaced by either a line connecting the two valid depth values at the boundary of the segment or a line with constant depth values at the boundaries of the image (see Figure 3). For textured planar surfaces (such as the checkerboard, see Figure 4(b)) this provides a reasonable estimate; if objects of low reflectance differ in depth from the surrounding they will be assigned wrong depth values, which will be corrected in the stereo part of the algorithm.

4   Algorithm

In our exemplary AR applications with dynamic occlusions we need to enhance one color image from the stereo camera with depth information. We compute depth values at the resolution of the PMD camera. The algorithm we suggest is equally applicable for computing depth at higher resolution or textured surfaces in other views.

The main steps of assigning depth values to pixels are as follows (see also Figure 5):

1. The pixel coordinates and depth values from the PMD camera are used for generating a tessellated depth surface (i.e. quad mesh) of the scene from this viewpoint.
2. The surface is transformed into the view space of the cameras. The intersections of view rays with the surface in this coordinate system define initial disparity values; the associated confidence values define the possible range. Thresholding the confidence values yields a set of valid and invalid depth coordinates in the quad mesh.
3. The areas of pixels associated to valid and invalid depth coordinates are thinned and serve as the initialization of a segmentation of the color image into depth continuous regions.

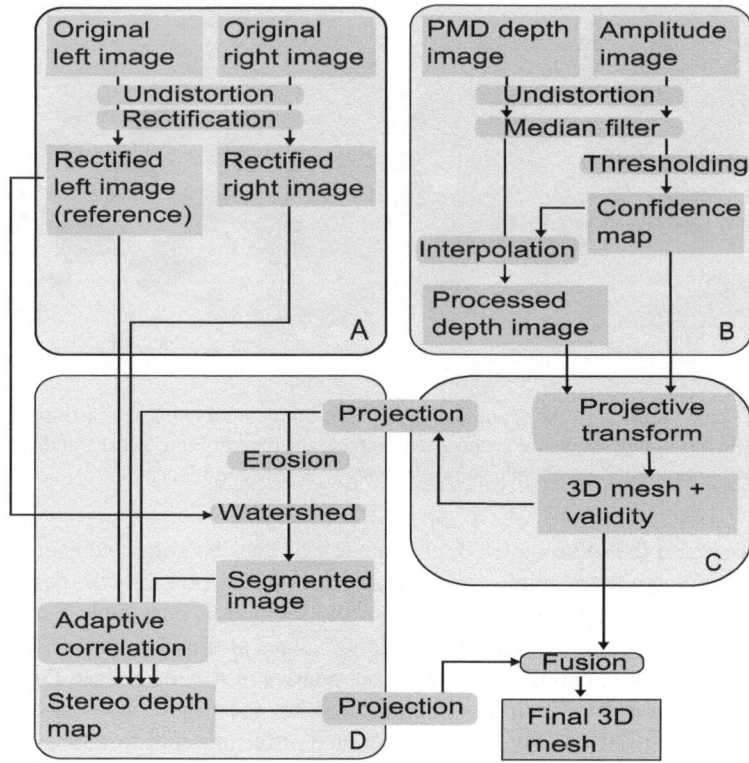

Fig. 5. Algorithm overview

4. The segmentation steers adaptive windows for the correlation computation in a standard stereo algorithm, correcting the invalid depth coordinates of the surface mesh.

In the following, we discuss several details of these steps.

4.1 Mesh Initialization and Projection

The intrinsic calibration of the PMD camera allows computing 3D coordinates from pixel location in the image plane and the corresponding depth value. For convenience, we connect the 3D coordinates to a piecewise bilinear mesh. The extrinsic calibration between the cameras allows transforming the mesh into the coordinate systems of the stereo camera. The intrinsic calibration of the stereo cameras (including a rectification) defines a projective transformation, which yields depth and confidence values per pixel.

The labels define two regions in the color images: a region of valid and a region of invalid pixels, based on the binary confidence map. Figure 6 shows the projection of the mesh in the left camera. The projection of valid vertices are drawn as gray squares and invalid pixels are drawn as red squares.

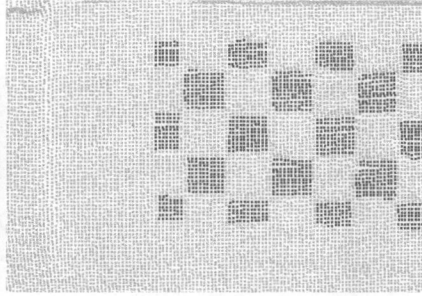

(a) Rectified Left View   (b) Stereo operation map

Fig. 6. Color coded vertices of the surface mesh, where red vertices are invalid and will be corrected using stereo vision

4.2 On-Demand Stereo with Adaptive Windows

Instead of computing a whole disparity map, the use of our stereo part is computing depth values only for vertices that are marked invalid. Furthermore, the projection of these vertices into the two rectified stereo views immediately yields an initial disparity guess.

An underlying assumption of correlation based stereo algorithms is that depth is unambiguous in the correlation window. This is not the case at depth discontinuities where objects may be occluded in only one of the views so that correlation of pixel colors fails to be a good indicator for correspondence (Figure 7(b) illustrates this situation).

A solution to this problem is to adapt the correspondence window to the (likely) object boundaries. Kanade and Okutomi [KO94] suggest to adapt the size and shape of the rectangular correlation window to local disparity characteristics. Boykov et al. generalize this variable window approach [BVZ98]. They compute for each pixel a new window. This window contains all pixels with an intensity close to the considered pixel. This way, they try to model the boundaries of the objects and the depth discontinuities. Hirschmüller [Hir01] proposes a similar approach using multiple supporting windows. Unfortunately, all of these techniques are too costly to reach interactive rates at video resolution of 640 by 480 pixels.

Our main observation is that the object boundaries are only relevant if they are in regions whose depth values are labeled as invalid – otherwise the depth values have already been gathered based on TOF. Thus, we can use the information on valid and invalid regions for initializing a segmentation algorithm in the color images. The segmentation will then define the extent of the correlation windows used in our adaptive window stereo algorithm. Exploiting the confidence information makes our approach both much faster and also more robust than only working with the color images.

From the many potential segmentation algorithms we use the *marker-controlled watershed* algorithm [RM00], which we have found to be robust while being fast enough for our application scenario. The idea is that valid and invalid regions serve as markers for the binary segmentation. Because of errors in the projections for vertices with incorrect depth (i.e. especially invalid vertices), color pixels are not necessarily labeled correctly. Consequently, the sets of valid and invalid pixels are eroded independently,

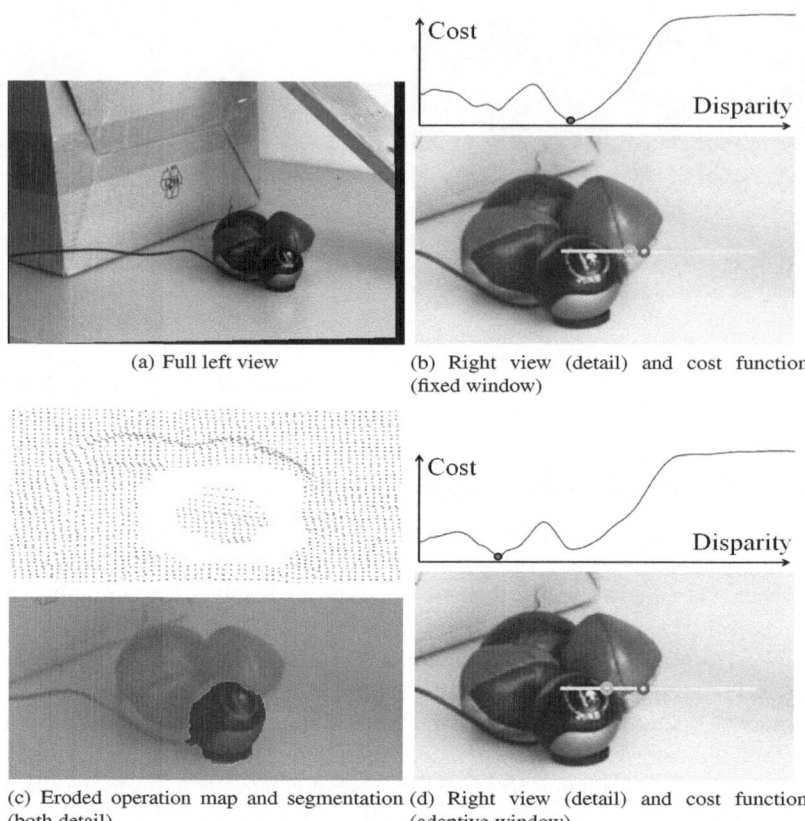

(a) Full left view

(b) Right view (detail) and cost function (fixed window)

(c) Eroded operation map and segmentation (both detail)

(d) Right view (detail) and cost function (adaptive window)

Fig. 7. This figure compares correlation based stereo with fixed windows and with windows adapted to object boundaries computed from segmenting the color image into depth continuous regions. The whole scene is shown (a), while we focus on the group of balls and the webcam in front of the box (b-d). The eroded operation map is used to initialize the watershed segmentation (c) leading to a mask adapting the stereo correlation windows. The red circle shows the initial disparity guess and the green circle the disparity corresponding to minimum cost (b+d).

leaving a set of unlabeled pixels in the proximity of object boundaries (see Figure 7(c)). These sets of valid and invalid pixels serve as the markers that initialize the segmentation as starting points. If objects have boundaries in the color images, the segmentation will accurately label pixels as being connected to the valid or invalid pixels. The resulting binary map restricts the correlation window.

Figure 7 shows the influence of this border correction filter: an object is too dark for the PMD camera, yielding wrong depth values and marked as invalid. A correlation based stereo algorithm with fixed window finds the wrong corresponding point (7(b)). After eroding the sets of invalid and valid pixels, the watershed algorithm segments the object along its boundary (7(c)). Restricting the window to the segmented object yields the correct correspondence (7(d)).

Table 1. Performance on an AMD Athlon 2.00 GHz Dual Core Processor with 1GB RAM. Window is the size of the correlation window used for stereo. The computation times are compared for each step while capturing a scene with approx. 250 and 500 corrected pixels.

Acquisition		91 ms	
Preprocessing		23 ms	
Stereo+PMD	Window	$\approx$ 250 px	$\approx$ 500 px
	5×5	**9.5 ms**	20.5 ms
	7×7	16 ms	33.3 ms
	9×9	25 ms	55 ms
	11×11	42 ms	88 ms

sensor types. In particular, we use range data for narrowing the disparity search and adapting the correlation windows to potential depth discontinuities.

We demonstrate how the resulting system can be used for handling occlusions. The depth data can also be used for collision detection and other interactive solutions. Some of the possible interactions are demonstrated in the accompanying video [htt09]. The dynamic and global depth map of the scene would also allow computing shadows for virtual objects, or using higher quality rendering techniques for further improving the realism of virtual objects [BR05].

Our system can be improved in several aspects. The synchronization of the PMD camera as part of the software system results in a maximum of 11fps – using a hardware solution would allow exploiting the maximum frame rate of the cameras. The computations necessary for fusing the data and improving the depth images are easy to distribute to several cores so that exploiting a higher input frame rate would be easy if the system were coupled with a modern CPU. As mentioned in section 2.2, the accuracy is depending on the PMD range data, and we think that future developments in camera hardware and calibration would lead to an increased working range.

It would be interesting to apply different segmentation algorithms for adapting the stereo windows. Feris et al. [FRC$^+$05] use region growing in a similar situation. Finding the best balance between performance and accuracy in this step is important future work. In addition, it might be possible to make use of other information than confidence and color, such as range data or predictions from preceding frames. Another step in enhancing the algorithm would be to define a confidence measure for the stereo data and use it for further controlling the depth reconstruction. This could reduce the error in regions where both systems fail, for example large dark and homogeneous objects.

References

[BBK07]   Beder, C., Bartczak, B., Koch, R.: A combined approach for estimating patchlets from PMD depth images and stereo intensity images. In: Hamprecht, F.A., Schnörr, C., Jähne, B. (eds.) DAGM 2007. LNCS, vol. 4713, pp. 11–20. Springer, Heidelberg (2007)

[BR05]   Bimber, O., Raskar, R.: Spatial Augmented Reality. A K Peters, Ltd., Wellesley (2005)

[BVZ98]   Boykov, Y., Veksler, O., Zabith, R.: A variable window approach to early vision. IEEE Trans. Pattern Anal. Mach. Intell. 20(12), 1283–1294 (1998)

[BWRT96]   Breen, D.E., Whitaker, R.T., Rose, E., Tuceryan, M.: Interactive occlusion and automatic object placement for augmented reality. Computer Graphics Forum 15(3), 11–22 (1996)

[FHM⁺93]   Faugeras, O., Hotz, B., Mathieu, H., Viéville, T., Zhang, Z., Fua, P., Théron, E., Moll, L., Berry, G., Vuillemin, J., Bertin, P., Proy, C.: Real time correlation based stereo: algorithm implementations and applications. Technical Report RR-2013, INRIA (1993)

[FHS07]   Fischer, J., Huhle, B., Schilling, A.: Using time-of-flight range data for occlusion handling in augmented reality. In: Eurographics Symposium on Virtual Environments (EGVE), pp. 109–116 (2007)

[FRC⁺05]   Feris, R., Raskar, R., Chen, L., Tan, K., Turk, M.: Discontinuity preserving stereo with small baseline multi-flash illumination. In: IEEE International Conference in Computer Vision (ICCV 2005), Beijing, China (2005)

[GAL07]   S. Guðmundsson, S., Aanæs, H., Larsen, R.: Fusion of stereo vision and time-of-flight imaging for improved 3D estimation. In: International workshop in Conjuction with DAGM 2007: Dynamic 3D Imaging, September 2007, vol. 1, pp. 164–172 (2007)

[HA07]   Hahne, U., Alexa, M.: Combining time-of-flight depth and stereo images without accurate extrinsic calibration. In: International workshop on Dynamic 3D Imaging, Heidelberg, September 2007, pp. 78–85 (2007)

[Hir01]   Hirschmüller, H.: Improvements in real-time correlation-based stereo vision. In: SMBV 2001: Proceedings of the IEEE Workshop on Stereo and Multi-Baseline Vision (SMBV 2001), Washington, DC, USA, p. 141. IEEE Computer Society, Los Alamitos (2001)

[htt09]   http://www.cg.tu-berlin.de/fileadmin/fg144/Research/ Publications/video/dyn3d09.mp4 (June 2009)

[KO94]   Kanade, T., Okutomi, M.: A stereo matching algorithm with an adaptive window: Theory and experiment. IEEE Trans. Pattern Anal. Mach. Intell. 16(9), 920–932 (1994)

[KOTY00]   Kanbara, M., Okuma, T., Takemura, H., Yokoya, N.: A stereoscopic video see-through augmented reality system based on real-time vision-based registration. In: Proceedings. IEEE Virtual Reality, 2000, pp. 255–262 (2000)

[KS06]   Kuhnert, K.-D., Stommel, M.: Fusion of stereo-camera and pmd-camera data for real-time suited precise 3d environment reconstruction. In: IEEE/RSJ International Conference on Intelligent Robots and Systems, October 2006, pp. 4780–4785 (2006)

[LK06]   Lindner, M., Kolb, A.: Lateral and depth calibration of PMD-distance sensors. In: Bebis, G., Boyle, R., Parvin, B., Koracin, D., Remagnino, P., Nefian, A., Meenakshisundaram, G., Pascucci, V., Zara, J., Molineros, J., Theisel, H., Malzbender, T. (eds.) ISVC 2006. LNCS, vol. 4292, pp. 524–533. Springer, Heidelberg (2006)

[LKH07]   Lindner, M., Kolb, A., Hartmann, K.: Data-fusion of pmd-based distance-information and high-resolution rgb-images. In: International Symposium on Signals, Circuits and Systems (ISSCS), Iasi, Romania (2007)

[LLK07]   Lindner, M., Lambers, M., Kolb, A.: Sub-pixel data fusion and edge-enhanced distance refinement for 2d/3d images. In: Dynamic 3D Imaging (Workshop in Cunjunction with DAGM 2007), Heidelberg, Germany (September 2007)

[MKF⁺05]   Moeller, T., Kraft, H., Frey, J., Albrecht, M., Lange, R.: Robust 3d measurement with pmd sensors. Technical report, PMDTec (2005)

[NMCR08] Netramai, C., Melnychuk, O., Chanin, J., Roth, H.: Combining pmd and stereo camera for motion estimation of a mobile robot. In: The 17th IFAC World Congress (July 2008) (accepted)

[Rap07] Rapp., H.: Experimental and theoretical investigation of correlating tof-camera systems. In: Physics, Faculty for Physics and Astronomy, University of Heidelberg, Germany (September 2007)

[Reu06] Reulke, R.: Combination of distance data with high resolution images. In: ISPRS Commission V Symposium 'Image Engineering and Vision Metrology' (2006)

[RM00] Roerdink, J.B.T.M., Meijster, A.: The watershed transform: Definitions, algorithms and parallelization strategies. FUNDINF: Fundamenta Informatica 41 (2000)

[SNV02] Schmidt, J., Niemann, H., Vogt, S.: Dense disparity maps in real-time with an application to augmented reality. In: WACV 2002: Proceedings of the Sixth IEEE Workshop on Applications of Computer Vision, Washington, DC, USA, p. 225. IEEE Computer Society, Los Alamitos (2002)

[SS02] Scharstein, D., Szeliski, R.: A taxonomy and evaluation of dense two-frame stereo correspondence algorithms. Int. J. Comput. Vision 47(1-3), 7–42 (2002)

[VH08] Ventura, J., Höllerer, T.: Depth compositing for augmented reality. In: SIGGRAPH 2008: ACM SIGGRAPH 2008 posters, p. 1. ACM, New York (2008)

[XSH+05] Xu, Z., Schwarte, R., Heinol, H., Buxbaum, B., Ringbeck, T.: Smart pixel - photonic mixer device (pmd) new system concept of a 3d-imaging camera-on-a-chip. Technical report, PMDTec (2005)

[Zha99] Zhang, Z.: A flexible new technique for camera calibration. Technical report, Microsoft Research (1999)

[Zha00] Zhang, Z.: A flexible new technique for camera calibration. IEEE Transactions on Pattern Analysis and Machine Intelligence 22(11), 1330–1334 (2000)

[ZWYD08] Zhu, J., Wang, L., Yang, R., Davis, J.: Fusion of time-of-flight depth and stereo for high accuracy depth maps. In: IEEE Computer Society Conference on Computer Vision and Pattern Recognition, CVPR (2008)

Realistic Depth Blur for Images with Range Data

Benjamin Huhle, Timo Schairer, Philipp Jenke, and Wolfgang Straßer

University of Tübingen, WSI/GRIS, Tübingen, Germany
{huhle,schairer,jenke,strasser}@gris.uni-tuebingen.de

Abstract. We present a system that allows for changing the major camera parameters after the acquisition of an image. Using the high dynamic range composition technique and additional range information captured with a small and low-cost time-of-flight camera, our setup enables us to set the main parameters of a virtual camera system and to compute the resulting image. Hence, the aperture size and shape, exposure time, as well as the focus can be changed in a postprocessing step. Since the depth-of-field computation is sensitive to proper range data, it is essential to process the color and depth data in an integrated manner. We use a non-local filtering approach to denoise and upsample the range data. The same technique is used to infer missing information regarding depth and color which occur due to the parallax between both cameras as well as due to the lens camera model that we use to simulate the depth of field in a physically correct way.

1 Introduction and Related Work

When photographing certain scenes it is a common stylistic device to limit the depth-of-field (DoF) in order to accentuate the actual motive and to provide a sense of depth within the scene. However, the artistic license of the photographer is often constrained by the capacities of the used camera. Especially with standard compact cameras the described technique employing a small DoF is only applicable in settings with very distant foreground and background objects. Often the aperture of the lens is not sufficient to achieve the desired effects. But even with high-end SLR cameras, the physical constraints remain valid and if the distances between fore- and background are not big enough, the whole picture remains sharp and the scene appears flat. Even in settings where the camera and the captured scene are suitable, it is often desirable to change the focus afterwards.

Different algorithms allow for the manipulation of images with respect to the mentioned degrees of freedom after the acquisition. In their recent work, Bae and Durand [1] detect depth blur in a single image and estimate the blur kernel in order to then magnify the amount of blur to achieve the desired DoF effect. Changing the focus, however, is not possible. Other approaches use a *depth from focus/defocus* technique (*e.g.* [19,15]). One inherent drawback of these methods is that the depth estimation is inaccurate depending on the surface texture. More

R. Koch and A. Kolb (Eds.): Dyn3D 2009, LNCS 5742, pp. 84–95, 2009.

related to our work is the technique by Moreno-Noguer *et al.* [16] who change the focus setting for an all-focused image in a post-processing step. A light pattern is projected into the scene from which they estimate sparse depth information. Based on image segmentation into regions of equal depth, the blur effect can be simulated. However, it is not possible to, *e.g.*, achieve a gradual increase in blur on surfaces parallel to the viewing direction.

In contrast, with the proposed setup consisting of a standard color camera and an additional range camera we acquire dense depth information in addition to the color image. We assume that the color image is all focused – a prerequisite that can be met by using a small aperture diameter, *i.e.*, a large *f*-number. The depth camera illuminates the scene with invisible near-infrared light and determines the time of flight of the reflected light. Since we obtain a dense depth map from the sensors it is possible to compute a gradual increase in blur on surfaces parallel to the viewing direction, for example.

Our processing pipeline incorporates a denoising stage, extending the recently presented non-local approach [12]. We first prune outlier pixels and secondly fuse color and depth information in order to enhance the resolution and smooth the depth map in an integrated manner.

Given correct depth information the depth blur can be simulated using different techniques. A good overview is given in [9] and we will use the categorization established there. The classical approach is the so-called *forward-mapped z-buffer DoF* where the circle of confusion is calculated for each pixel and rendered to the destination image [17]. The result is equivalent to using the *reverse-mapped z-buffer DoF* technique where a blur filter with a kernel size depending on the corresponding depth value is applied to the image. Both techniques do not take occlusions into account, *i.e.* no depth test is performed. Therefore blurred background objects can "leak" into the foreground. The *layered DoF* algorithm assigns pixels of the image to distinct "slices" of depth that can be blurred based on some representative depth. These layers are composited into a final image and in this way occlusions between these layers are handled correctly. Moreno-Noguer *et al.* [16] apply this technique when performing refocusing.

The exact way for computing DoF is to model the light transport from the scene to the camera. This is done in distributed ray tracing by casting rays from across the lens [5]. Equivalently, the scene can be rendered multiple times from different locations on the lens and accumulating the resulting images [11,4]. This is a correct reproduction of the depth of field effect according to a thin lens camera model. However, for the simulation, knowledge about the scene geometry is necessary. Since every point on a real lens sees different parts of the scene, a single depth map is not sufficient. Therefore, missing structure due to occlusions in the depth map has to be inferred. Our novel approach is to cast this 3D problem into the image domain.

The presented technique uses an inpainting algorithm for both color and depth in order to achieve the realistic depth blur. The inpainting problem was studied intensively in the last years. In the seminal paper by Efros and Leung [8] missing pixels are reconstructed using pixels with a similar local neighborhood.

Extensions were presented *e.g.* by Wei and Levoy [21]. We use the same non-local approach employing statistics about the set of patches in the depth map and the image as is used for denoising. In an integrated manner, these patch statistics are used to infer missing pixels that occur due to the small parallax effect between different viewpoints on the lens, similar to the work by Wong and Orchard [22].

The depth images we acquire with the time-of-flight (ToF) camera are of low resolution and have to be upsampled to the resolution of the color image. Inferring high resolution versions from a given low resolution image is also a long studied topic and several approaches based on learned dependencies of low and high resolution example data have been presented (*e.g.* [10,13]). Here, however, we are dealing with the super-resolution problem in a sensor fusion framework where both color and depth are available. We integrate the resolution enhancement into the same non-local technique that is used for denoising. Related to ours are the approaches based on an extended Markov random field [7] and on the joint bilateral filter [14].

By varying the exposure time, *e.g.*, with a bracket mode setting, and applying Debevec and Malik's composition technique [6] our algorithm correctly computes the depth blur on the irradiance values of high dynamic range. Therefore, we are able to reproduce a realistic *Bokeh*. This term describes the visual quality of the depth blur effect and is often associated with the occurrence of single highlighted spots that are characteristic for a certain shape of the aperture. Commercial tools, *e.g.* Photoshop's lens blur filter, only simulate this effect by "boosting" bright areas in 8bit-images.

In the following section we give a brief overview over the system and explain the individual steps of our approach in detail. Results from different test scenes are discussed in Section 3 and we conclude the paper in the last section.

2   Our Approach to Refocusing

2.1   Overview

For the acquisition of a color image with corresponding depth information we use a setup consisting of a 2 Megapixel standard color camera that is mounted on top of a time-of-flight camera by *PMDTec*, resulting in a parallax of about 50 mm. The range camera provides depth maps with a resolution of 160×120 pixels and an accuracy that highly depends on the scene and the given lighting conditions. Both cameras are calibrated once with a standard camera calibration toolkit [2] using the intensity output of the range camera, equivalently to a stereo setup. Based on the intrinsic calibration, the color images are undistorted prior to further processing. To circumvent any interpolation of the depth data we compensate for the distortion of the lens by incorporating these effects in the backprojection using a precomputed lookup table. An overview of our processing pipeline is depicted in Figure 1 and explained in the remainder of this section.

Capturing a scene, we take images of the color camera using several exposure times together with a range image from the ToF camera. In order to obtain an all-focused image, a small aperture (big f-number) has to be used.

In a postprocessing stage, the color images are composited to a high dynamic range image using the algorithm developed by Debevec and Malik [6]. Outliers in the depth data, *i.e.*, pixels where the depth measurement failed due to illumination conditions or reflection properties, are pruned with a non-local filter [12]. Further, we backproject the depth readings into 3D-space and project them onto the image plane of the color camera. Due to occlusions and the low resolution of the range camera, we need to upsample and inpaint the depth map in order to obtain corresponding range measurements for each image pixel. With a novel unified approach based on a non-local filter the depth image is denoised at the same time. We describe this filter in Section 2.2.

The refocusing and simulation of the DoF effect is achieved by integrating over all possible viewpoints on the lens within the chosen aperture. This is done by stochastic sampling, *i.e.*, rendering the scene from several randomly chosen viewpoints and accumulating the views. When the viewpoint changes on the lens, this small parallax effect results in holes within the image. Therefore, each rendering pass is followed by an inpainting step. This DoF simulation is described in detail in Section 2.3.

2.2   Super-Resolution and Inpainting of Depth

Buades recently presented the *Non-Local Means* (NL-means) filter for image restoration [3]. It reconstructs pixels using other pixels with similar local neighborhoods, taking into account the self-similarity of the image. Even very fine details that occur repeatedly can be distinguished from noise. In this manner, NL-means is also closely related to the inpainting algorithm by Efros and Leung [8]. In its original formulation NL-means reconstructs a pixel by a weighted average

$$v'(\mathbf{i}) = \frac{1}{Z_{\mathbf{i}}} \sum_{\mathbf{j} \in \mathbf{W_i}} w(\mathbf{i}, \mathbf{j}) v(\mathbf{j}). \tag{1}$$

of pixels $v(\mathbf{j})$ that share a similar surrounding $\mathbf{N}$. Here, $Z_{\mathbf{i}} = \sum_{\mathbf{j} \in \mathbf{W_i}} w(\mathbf{i}, \mathbf{j})$ denotes a normalization constant and $\mathbf{W_i}$ is a potentially large search window around the 2D position $\mathbf{i}$. The similarity weight w is determined as

$$w(\mathbf{i}, \mathbf{j}) = e^{-\frac{1}{h} \sum_{\mathbf{k} \in \mathbf{N}} \xi_{\mathbf{ik}} G_a(\|\mathbf{k}\|_2)(v(\mathbf{i}+\mathbf{k}) - v(\mathbf{j}+\mathbf{k}))^2}, \tag{2}$$

with filtering parameter h. The pixel-wise distances are weighted according to their offset $\mathbf{k}$ from the central pixel using a Gaussian kernel G_a with standard deviation a. In this work we use a variant (see [12]), where the additional factor

$$\xi_{\mathbf{ik}} = e^{-\frac{(v(\mathbf{i}) - v(\mathbf{i}+\mathbf{k}))^2}{h}} \tag{3}$$

is used to ensure a proper smoothing along strong discontinuities and therefore accounts for the different characteristics of typical depth maps compared to images. Outliers are pruned using the conditional probability density $p\left(v(\mathbf{i})|v(\mathbf{N_i^*})\right)$

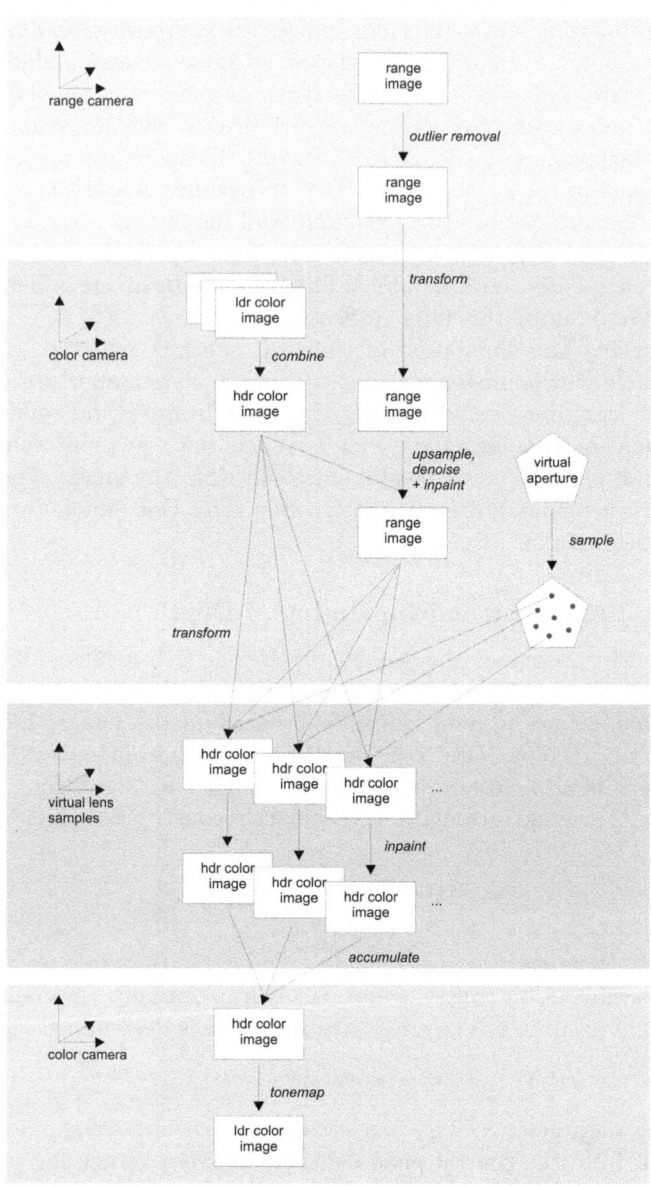

Fig. 1. Pipeline of the proposed system

modeling the data with a mixture of a normal distribution for valid pixels and a uniform distribution for outlier pixels according to [12], where $\mathbf{N_i^*} = \mathbf{N} \setminus \mathbf{i}$. In our setting the outlier removal is performed prior to the transformation into the color camera frame in order to avoid false occlusions.

Once depth and image information is given in the same reference frame we smooth and upsample the depth data since depth and color are available in different resolutions. Using a linear blending we reconstruct the depth values on the fine grid as

$$v'_{uv}(\mathbf{i}) = \frac{1}{Z_\mathbf{j}^*} \sum_{\mathbf{j} \in \mathbf{W_i}} \Big((1 - \alpha)\, w(\mathbf{i}_\downarrow, \mathbf{j}_\downarrow) + \alpha w^{(u)}(\mathbf{i}, \mathbf{j}) \Big)\, v(\mathbf{j}_\downarrow), \qquad (4)$$

where $Z_\mathbf{j}^*$ denotes the corresponding normalization factor. Here, $\mathbf{i}_\downarrow$ are the low resolution pixels that correspond to pixels $\mathbf{i}$ in the high resolution grid. The image based weights $w^{(u)}$ are defined analogously to the similarity weights w. Whereas the size of the neighborhoods $\mathbf{N}$ for the computation of $w^{(u)}$ is related to the high resolution image, the neighborhood used in the calculation of w has to be chosen according to the low resolution depth data.

Individual pixels that were marked outliers in the prior detection step as well as holes resulting from the transformation between both camera frames can be reconstructed in the same way. In order to fill missing areas that exceed the patchsize, the filter is applied iteratively. For each reconstructed pixel we use $Z_\mathbf{j}^*$ as a measure of certainty $\zeta(\mathbf{j})$ for the next iteration. Initially, $\zeta(\mathbf{j})$ is set to 1 for valid pixels and 0 otherwise. Equation 2 is modified in order to propagate the confidence measure into areas of missing pixels:

$$w(\mathbf{i}, \mathbf{j}) = \zeta(\mathbf{j}) e^{-\frac{1}{h} \sum_{\mathbf{k} \in \mathbf{N}} \xi_{\mathbf{i}\mathbf{k}} G_a(\|\mathbf{k}\|_2) \zeta(\mathbf{i}) \zeta(\mathbf{j}) (v(\mathbf{i}+\mathbf{k}) - v(\mathbf{j}+\mathbf{k}))^2}. \qquad (5)$$

Iteratively computing the reconstruction (Eq. 1) for each pixel and storing the corresponding confidence measures, the depth data can be upsampled to the resolution of the image data and smoothed laterally as well as in the range dimension in one unified approach.

2.3   Computing Realistic Depth Blur

To achieve the desired depth-of-field effect, one has to (at least preliminarily) cede the widely used pinhole camera model. Instead, we have to follow the thin lens camera model incorporating the aperture size as an additional parameter.

A direct implementation of this model, however, is only possible if the scene and the occlusions in 3D space are known. Since a single depth map is the only information about the scene geometry in our setting, we compensate for the problem of unknown parts of the scene by simulating the thin lens mapping using several pinhole projections and inpainting resulting holes in the image domain on each of them. The same inpainting technique as described in the previous section is used. Holes that occur due to a changed viewpoint are most probably to be filled extending the adjacent background. Therefore, a prior is

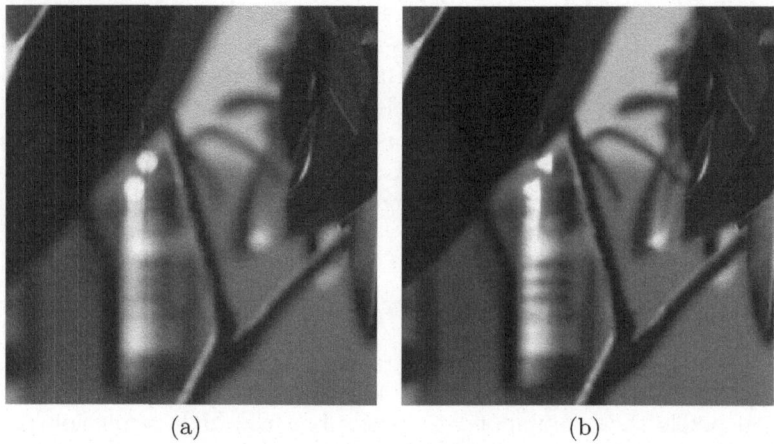

Fig. 2. Bokeh: Two different aperture shapes; 6-blade (a) and triangle-shaped (b)

added in Equation 5 that assigns a higher weight to patches of larger depth. We scale the weight $w(\mathbf{i}, \mathbf{j})$ with the corresponding squared depth of the pixel.

In a first step we sample 2D positions $s = (s_x, s_y)$ on the lens of a virtual camera using stratified sampling [20,18]. Given the focal length f and the f-number N chosen by the user, we sample from a disc with diameter $c = f/N$ that represents the unoccluded part of the lens. Different DoF effects can easily be achieved by sampling from varying aperture shapes instead.

The pinhole projection for each sample is applied after a transformation of the camera frame, translating the viewpoint according to the 2D displacement on the lens (perpendicular to the viewing direction) and perturbating the viewfrustum such that the image coordinates of points at a chosen distance u (the focus) remain unchanged. This is done by shearing the scene geometry parallel to the image plane.

3   Results

We implemented a prototype version of the presented system in C++, performing the filtering steps on the GPU. For all the examples in the paper we use 256 samples (576 for the close-ups) to integrate over the lens (see Sec. 2.3). As target resolution we use an image size of up to 1600×1200 pixels, *i.e.*, we upsample the depth data by a factor of 10 in both dimensions.

A scene of complex depth structure is shown in Figure 3a-d. Compared to the unmodified image (a), the depth perception is significantly enhanced by introducing depth blur. Using the presented framework we varied the focus to three different stages: close (b), medium (c) and far (d). At the highlights on the fire extinguisher one can clearly observe the desired bokeh effect. In the close-up view in Figure 2 we show the resulting effect using different aperture shapes. The effect of photometric burnout can be noticed along the branch in the marked area in (b).

(a) original   (b) focus: 1.5m

(c) focus: 2.0m   (d) focus: 3.8m

Fig. 3. *Rubber Tree Scene*: Refocused with a hexagonal (6-blade) aperture

The second scene depicted in Figure 4 shows how our system handles gradual changes in depth (in contrast to the system presented in [16]). Please note the continuously increasing amount of depth blur on the checkered table cloth. Additionally, we show different exposure times and their varying bokeh effects.

A compariso with the commonly used reverse-mapped z-buffer DoF (see Section 1) is shown in Figure 5 using a texture-less synthetic scene. Here, some disturbing artifacts of the reverse-mapping approach can be clearly noticed. Most prominently, the edge of the blurred red cylinder remains sharp; this is because the green cylinder (near the focal plane) remains sharp and each pixel representing the green cylinder therefore is assigned a high accumulation weight. Correspondingly, a step artifact occurs where the background of the red cylinder changes from sharp (the green cylinder) to the blurry background with low accumulation weight (detail-view in Subfigure (c)). Using the proposed pipeline, a physically correct photometric burnout occurs and the step artifact at the changing background (detail-view in Subfigure (d)) is reduced.

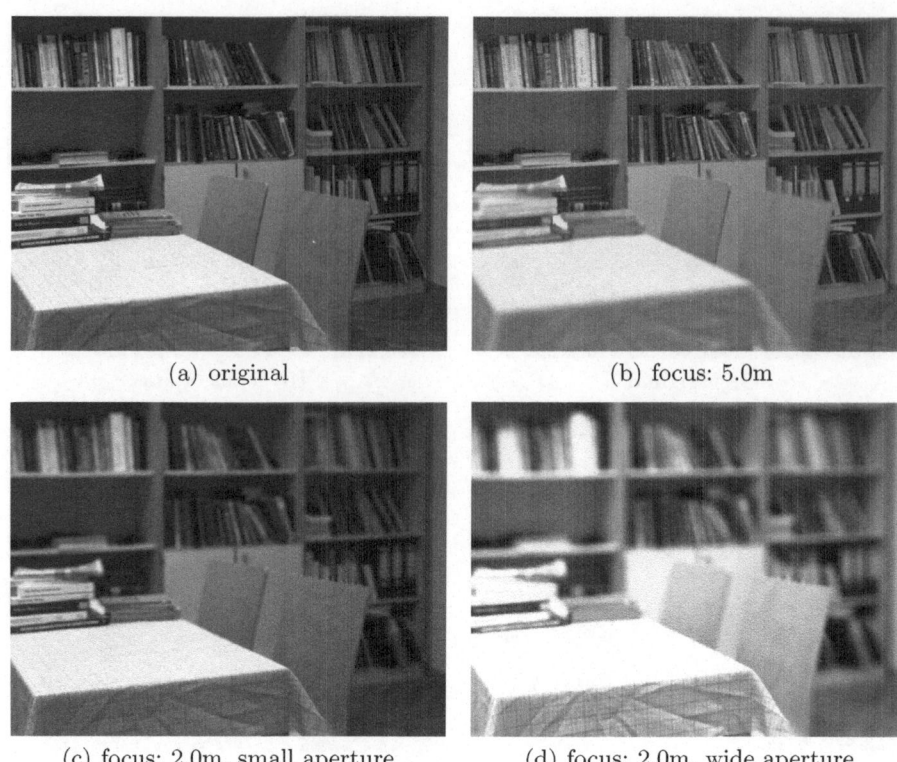

(a) original   (b) focus: 5.0m

(c) focus: 2.0m, small aperture   (d) focus: 2.0m, wide aperture

Fig. 4. *Table Scene*: Refocused with a round aperture; varied exposure settings

An important prerequisite for the proposed system is a correct alignment of depth and color data. A proper calibration of both cameras is therefore essential. Furthermore, in our experiments we experienced mainly two sources of error. First, depending on the resolution of the depth sensor and its accuracy, erroneous distance measurements might be fed into the pipeline. This can lead to the effect, that parts of the image are blurred whereas they should be in focus and vice versa. Also, the parallax effects might not be compensated exactly when transferring the depth data to the color reference frame or when changing the view point on the lens. Secondly, in the inpainting and super-resolution step, the algorithm relies on the heuristic that discontinuities in depth and color tend to coalign. If this does not hold, wrong distance values might be assigned to the image pixels around edges in depth where the color is similar on both sides. Consequently, the focus is set incorrectly (see the backrest of the closer chair in Fig. 4). However, due to the similarity in color, this effect is not always noticeable.

More results in full resolution as well as a short video illustrating the pipeline can be found at http://www.gris.uni-tuebingen.de/people/staff/huhle/

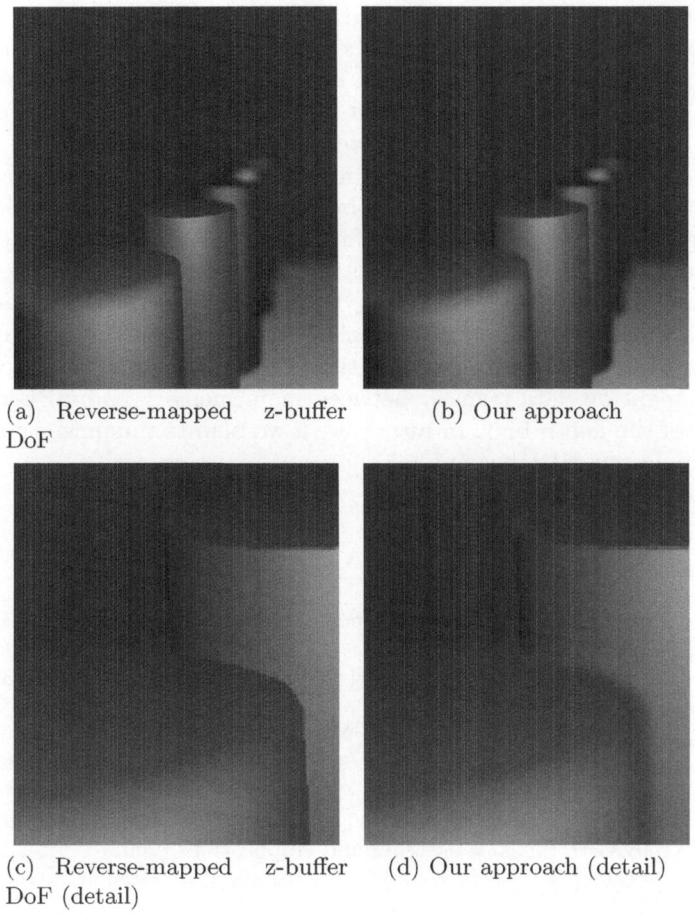

(a) Reverse-mapped   z-buffer   (b) Our approach
DoF

(c) Reverse-mapped   z-buffer   (d) Our approach (detail)
DoF (detail)

Fig. 5. *Synthetic Scene*: Comparison with common filtering approach. The camera is focused on the green cylinder.

4   Conclusion

The presented system and method allow a user to manipulate the major camera parameters (aperture shape and size, focus as well as exposure time) after acquisition. We simultaneously capture the scene with a color and a time-of-flight depth camera. Such range cameras could be manufactured efficiently and hence, this setup is prospectively also realizable even in price-sensitive systems. The contribution of our work consists of two main points, namely the preprocessing step and the actual computation of the depth blur. First, color and depth data are preprocessed in an integrated manner using a unified non-local filter for denoising, upsampling and inpainting of missing information that occurs due to parallax effects. Secondly, we simulate a thin lens model by sampling the shape of the aperture in order to compute the DoF effect. Employing this technique, the problem of handling occlusions in the scene is cast to inpainting in the image domain. The physically correct camera model enables us to set the parameters of our virtual camera exactly as with a real camera. In comparison with state-of-the-art techniques, our novel approach reduces the artifacts that occur due to limited knowledge of the scene geometry captured in a single depth map. However, the prototype system should be considered a proof of concept and the overall quality of the results will benefit from future improvements concerning each of the steps in the pipeline. In particular, more effort could be spent on the sampling issues when transforming depth and image data between different viewpoints. Regarding the implementation of the system, the individual filtering steps are performed efficiently on the GPU, however, a major part of the runtime is spent on data transfer between main memory and GPU during the generation of the depth blur. In future work we plan to minimize this overhead and to include several other optical effects of real cameras (*e.g.* spherical and chromatic abberations).

References

1. Bae, S., Durand, F.: Defocus Magnification. Computer Graphics Forum (Proc. EUROGRPHICS) 26, 571–579 (2007)
2. Bouguet, J.-Y.: Camera Calibration Toolbox for Matlab. California Institute of Technology
3. Buades, A., Coll, B., Morel, J.-M.: A Non-Local Algorithm for Image Denoising. In: Proc. IEEE Conference on Computer Vision and Pattern Recognition, CVPR (2005)
4. Buhler, J., Wexler, D.: A Phenomenological Model for Bokeh Rendering. In: Prof. ACM SIGGRAPH, Session: Sketches and Applications (2002)
5. Cook, R., Porter, T., Carpenter, L.: Distributed Ray Tracing. In: Proc. ACM SIGGRAPH (1984)
6. Debevec, P.E., Malik, J.: Recovering High Dynamic Range Radiance Maps from Photographs. In: Proc. ACM SIGGRAPH (1997)
7. Diebel, J., Thrun, S.: An Application of Markov Random Fields to Range Sensing. In: 19th Annual Conference on Neural Information Processing Systems, NIPS (2005)
8. Efros, A.A., Leung, T.K.: Texture Synthesis by Non-parametric Sampling. In: Proc. IEEE International Conference on Computer Vision, ICCV (1999)
9. Fernando, R.: GPU Gems: Programming Techniques, Tips and Tricks for Real-Time Graphics. Pearson Higher Education, London (2004)
10. Freeman, W.T., Jones, T.R., Pasztor, E.C.: Example-Based Super-Resolution. IEEE Computer Graphics and Applications 22(2), 56–65 (2002)
11. Haeberli, P., Akeley, K.: The Accumulation Buffer: Hardware Support for High-Quality Rendering. In: Proc. ACM SIGGRAPH (1990)
12. Huhle, B., Schairer, T., Jenke, P., Straßer, W.: Robust Non-Local Denoising of Colored Depth Data. In: Proc. IEEE CVPR Workshop on Time-of-Flight Based Computer Vision, TOF-CV (2008)
13. Kim, K.I., Franz, M.O., Schölkopf, B.: Iterative Kernel Principal Component Analysis for Image Modeling. IEEE Transactions on Pattern Analysis and Machine Intelligence (PAMI) 27(9), 1351–1366 (2005)

14. Kopf, J., Cohen, M.F., Lischinski, D., Uyttendaele, M.: Joint Bilateral Upsampling. In: Proc. ACM SIGGRAPH, New York, NY, USA, p. 96 (2007)
15. McGuire, M., Matusik, W., Pfister, H., Hughes, J.F., Durand, F.: Defocus Video Matting. In: Proc. ACM SIGGRAPH, vol. 24, pp. 567–576. ACM Press, New York (2005)
16. Moreno-Noguer, F., Belhumeur, P.N., Nayar, S.K.: Active Refocusing of Images and Videos. In: Proc. ACM SIGGRAPH (2007)
17. Potmesil, M., Chakravarty, I.: A Lens and Aperture Camera Model For Synthetic Image Generation. In: Proc. ACM SIGGRAPH (1981)
18. Shirley, P.: Discrepancy as a Quality Measure for Sampling Distributions. In: Computer Graphics Forum (Proc. EUROGRAPHICS), pp. 183–194 (1991)
19. Subbarao, M., Wei, T.-C., Surya, G.: Focused Image Recovery from Two Defocused Images Recorded with Different Camera Settings. IEEE Transactions on Image Processing 4, 1613–1628 (1995)
20. Turk, G.: Generating Random Points in Triangles. Academic Press Professional, Inc., San Diego (1990)
21. Wei, L.-Y., Levoy, M.: Fast texture synthesis using tree-structured vector quantization. In: Proc. ACM SIGGRAPH, New York, NY, USA, pp. 479–488. ACM Press/Addison-Wesley Publishing Co. (2000)
22. Wong, A., Orchard, J.: A Nonlocal-Means Approach to Exemplar-Based Inpainting. In: IEEE Int. Conf. on Image Processing (2008)

Global Context Extraction for Object Recognition Using a Combination of Range and Visual Features

Michael Kemmler, Erik Rodner, and Joachim Denzler

Chair for Computer Vision
Friedrich Schiller University of Jena
{Michael.Kemmler,Erik.Rodner,Joachim.Denzler}@uni-jena.de
http://www.inf-cv.uni-jena.de

Abstract. It has been highlighted by many researchers, that the use of context information as an additional cue for high-level object recognition is important to close the gap between human and computer vision. We present an approach to context extraction in the form of global features for place recognition. Based on an uncalibrated combination of range data of a time-of-flight (ToF) camera and images obtained from a visual sensor, our system is able to classify the environment in predefined places (e.g. kitchen, corridor, office) by representing the sensor data with various global features. Besides state-of-the-art feature types, such as power spectrum models and Gabor filters, we introduce histograms of surface normals as a new representation of range images. An evaluation with different classifiers shows the potential of range data from a ToF camera as an additional cue for this task.

1   Introduction

The development of time-of-flight (ToF) cameras [1], which provide range information in realtime, has led to a large number of applications. Most of them concentrate on the support of vision-based systems in tasks like 3D reconstruction and robot navigation [2]. Alternatively to geometric reconstruction techniques, we show how to utilize a classification based system for place recognition or rough self localization of a mobile robot.

Instead of describing the position of a robot in exact geometric terms, it is often beneficial to use a discretization of predefined places or scenes, e.g. kitchen, corridor or office. Especially for subsequent object detection tasks [3], information about the current place can be used as high-level contextual information [4]. Due to the large variability of scene appearances, the estimation of the most probable label is a challenging recognition task. For this reason we calculate a feature representation from ToF range data and from an image obtained using a standard visual sensor (Fig. 1). This allows to describe a scene using rough 3D information and visual appearance. Furthermore we present a simple method for feature calculation in range images which describes the image as a collection

R. Koch and A. Kolb (Eds.): Dyn3D 2009, LNCS 5742, pp. 96–109, 2009.

Fig. 1. Setup of our place recognition system with a ToF sensor and a visual sensor mounted on a mobile robot. Data is obtained from both uncalibrated cameras in order to build the combined feature representation of the current view.

of planar patches. It can be seen as an instance of the bag-of-features concept, which has been shown to be well suited for scene recognition [5]. Features from visual images are calculated using two state-of-the-art approaches often used for the task of scene recognition. Our work extends the scene recognition approach of [4] to multiple sensors and range data.

The remainder of the paper is organized as follows: First of all, we present histograms of surface normals as a feature type for range images which is well suited for the place recognition task. In Sect. 3 we describe state-of-the-art global feature representations that can be applied to data from the visual and the range sensor. Classification techniques and details of the feature combination are explained in Section 4. Experiments in Sect. 5 compare feature types and different classifiers and show the performance benefit of feature combination from different sensors. A summary of our findings and a discussion of future research directions conclude the paper.

2   Histogram of Surface Normals

Range images captured by ToF sensors consist of dense distance measurements of scene elements in the field of view of the camera. Using a simple histogram representation of all depth values would be a typical global representation of the scene. However, for scene and place recognition with standard cameras, feature types that use aggregated local statistics of pixel neighborhoods showed to be successful. A simple but efficient approach to incorporate information from a small environment of a pixel is the representation of a range image as a collection of small planar patches or patchlets [6]. A statistic of the orientation of such planar patches then corresponds to local surface characteristics.

Let $\boldsymbol{x}$ be a three dimensional point obtained from the range image and $N(\boldsymbol{x})$ the set of all points in the (rectangular) image neighborhood of size $P \times P$ with center $(\boldsymbol{x}_1, \boldsymbol{x}_2)^T$.

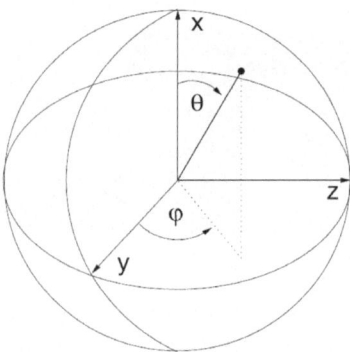

Fig. 2. Representation of surface normals in sphere coordinates [7]

In the following we assume orthogonal projection. Note that we will show that our scene recognition system achieves a suitable performance without the need for an intrinsic camera calibration. With given camera parameters one can easily undo the perspective projection, which might result in a better recognition performance. Nevertheless this influence is not investigated in this paper, because our results show that despite our severe assumption a histogram of surface normals can be a useful feature representation (cf. Sect. 5.2).

Each plane that does not intersect the camera center can be described by $n^T x = 1$, where $n = (n_x, n_y, n_z)^T$ denotes the surface normal. We estimate the parameters of the planar patch in each point x^i with Iteratively Reweighted Least Squares (IRLS) applied to the resulting optimization problem:

$$n^i = \arg\min_{n} \sum_{x \in N(x^i)} |n^T x - 1| \; . \tag{1}$$

Instead of absolute depth values, we use local surface characteristics as a feature. Therefore we utilize the normal representation of Hetzel et al. [7], which transforms n^i into a pair of angles $(\varphi^i, \theta^i)^T$ in sphere coordinates, where:

$$\varphi = \arctan\left(\frac{n_z}{n_y}\right) \tag{2}$$

$$\theta = \arctan\left(\frac{\sqrt{n_y^2 + n_z^2}}{n_x}\right) \tag{3}$$

as illustrated in Fig. 2. Thus, the resulting representation is a two dimensional histogram with B_φ and B_θ bins for ϕ^i and θ^i, and $B_\varphi \times B_\theta$ entries.

3   Visual Features

In the subsequent sections low-level visual features are described, which we utilize to calculate a feature representation of the data of our visual sensor. Additionally,

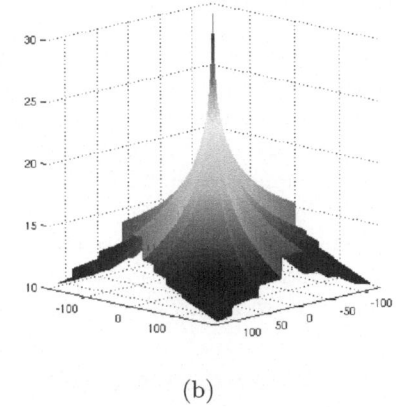

(a)   (b)

Fig. 3. Sample image (a) and its (logarithmed) power spectrum representation with 16 sectors (b)

we use the following features to extract second order and structure information from range images.

3.1 Power Spectrum Features

One famous approach, which was first described by Mezrich et al. [8] in the late seventies, is to fit the Fourier power spectrum to an isotropic model. Empirical studies on natural images [8,9] show that the average power spectrum approximately obeys the power law $M(\boldsymbol{f}) = A \cdot ||\boldsymbol{f}||_2^{-\alpha}$, with parameter A and α, where $\boldsymbol{f}$ denotes frequency. Straightforward linear least squares optimization can be used to estimate the model parameters.

However, Oliva and Torralba [9] empirically show that the power law does not hold for artificial images. Thus, since we concentrate on indoor environments and want to calculate features from a single image, it is unlikely that an isotropic representation is sufficient to properly describe present second order statistics. We therefore use an extended representation [9], where the power spectrum is radially divided in Ω non-overlapping sectors. Each sector ω is then assumed to obey a power law:

$$M_\omega(\boldsymbol{f}) = \frac{A_\omega}{||\boldsymbol{f}||_2^{\alpha_\omega}} \quad 1 \leq \omega \leq \Omega \; . \tag{4}$$

In order to reduce noise, radially averaging [10] is employed for each sector prior to model fitting. Note that this anisotropic power spectrum model, which is illustrated in Fig. 2 does not incorporate phase information.

In the remainder of this paper, a 16-sector model is used which results in a 32-dimensional feature vector $(\alpha_1, \ldots, \alpha_{16}, A_1, \ldots, A_{16})$.

3.2   Gabor Features

Phase-preserving representations can be computed using properties of the amplitude spectra. Gabor filters are selective filters that respond to structures of a specific range of frequencies and orientations. A bank of Gabor filters, therefore, can be used as a global image representation. Since the collection of responses is very high-dimensional, we follow the approach of [11], where subsampled squared response images are used. This results in substantially reduced feature vectors. Prior to Gabor filtering, the image is preprocessed by a whitening step, followed by divisive normalization [12] in order to increase contrast and, thus, amplify higher-order structures.

4   Classification and Feature Combination

In this paper, four different classifiers were used in order to learn the mapping between features and scene labels: multi-layer Perceptron (MLP), Parzen classifier, Randomized Decision Forests, and Support Vector Machines. However, for the sake of brevity, only the latter three classifiers are described here.

4.1   Parzen Classifier Using Kernel Density Estimation

Core of the generative Parzen classifier for Gaussian kernel densities [13,14] is the estimation of empirical likelihoods for each class $\kappa \in \{1, \ldots, K\}$:

$$p(\mathbf{f} \mid S_\kappa) = \frac{1}{M_\kappa} \sum_{i=1}^{M_\kappa} \mathcal{K}_\kappa(\mathbf{f} - \mathbf{f}_i) , \qquad (5)$$

where $\mathcal{K}_\kappa$ is a zero-mean normal density with covariance matrix $\boldsymbol{\Sigma}_\kappa$ and the set $S_\kappa = \{\mathbf{f}_1, \ldots, \mathbf{f}_{M_\kappa}\}$ denotes the n-dimensional training data labeled with class κ. An unseen feature $\mathbf{f}$ is then classified using maximum likelihood estimation.

Although the shape of the empirical density is determined by the observed data S_κ, the smoothness depends solely on the kernel bandwidth parameter $\boldsymbol{\Sigma}_\kappa$. The appropriate choice of a bandwidth is the most critical step in kernel density estimation, since small bandwidths lead to over-fitting, whereas too large bandwidths result in oversmooth densities. In this paper, we use an ad-hoc method for bandwidth selection known as generalized *Scott's rule* [14] for kernel densities:

$$\boldsymbol{\Sigma}_\kappa \approx M_\kappa^{-\frac{1}{n+4}} \widehat{\boldsymbol{\Sigma}}_\kappa^{\frac{1}{2}} , \qquad (6)$$

where $\widehat{\boldsymbol{\Sigma}}_\kappa$ is the sample covariance with respect to S_κ.

4.2   Randomized Decision Forest

A Randomized Decision Forest (RDF) is a discriminative classifier that can handle a large set of features without issues due to the curse of dimensionality. Standard decision tree approaches suffer from severe over-fitting problems.

A RDF overcomes these problems by generating an ensemble (forest) of T decision trees. During the classification, the overall probability of a class κ given a feature vector $\mathbf{f}$ can be obtained by simple averaging of the posterior probabilities $p_\tau(\cdot)$ estimated by each tree of the ensemble:

$$p(\kappa \mid \mathbf{f}) = \frac{1}{T} \sum_{\tau=1}^{T} p_\tau(\kappa \mid \mathbf{f}) \ . \tag{7}$$

In contrast to Boosting, the RDF approach uses two types of randomization to learn the ensemble. The first type of randomization is Bootstrap Aggregating [15], where each tree is trained with a random fraction of the training data. Additionally, to reduce training time and to incorporate randomization into the building process of a tree, the search for the most informative split function in each inner node is done using only a random fraction of all features [16].

4.3 Support Vector Machines

In the last years, Support Vector Machines (SVM) have emerged to one of the most popular machine learning techniques. For a basic introduction we refer the reader to the textbook of Bishop [13] and concentrate on the detailed setup used for our evaluation.

We train K SVM classifiers using the one-vs.-all principle. All scores are converted to suitable probabilities using the logistic regression method of Platt et al. [17]. The classification result is the class with the highest probability (score of the corresponding binary SVM classifier). Each single classifier uses a radial basis function kernel with parameter γ and trade-off parameter C [13] optimized with cross-validation. Instead of simple grid search, we apply cyclic coordinate search which is faster and yields in our experiments to similar optimal parameters.

4.4 Feature Combination and Temporal Context

In order to combine a set of features $\mathbf{F} = \{\mathbf{f}_1, \ldots, \mathbf{f}_{|\mathbf{F}|}\}$, simple concatenation is performed. To avoid facing the curse of dimensionality, which often occurs with generative classifiers, a different scheme is used for the Parzen classifier. In addition to subspace reduction via PCA, we choose a soft voting approach, where each feature type $\mathbf{f}_i$ is classified separately. The overall class probability $p(\kappa|\mathbf{F})$ is then computed by averaging the separate class probabilities $p(\kappa|\mathbf{f}_i)$.

To further improve the classification performance, a hidden Markov model (HMM) is used to exploit temporally contextual properties. We use the approach from Torralba et al. [4], but instead of a sparse Parzen classifier, we utilize the classifiers listed above.

5   Experiments

We experimentally evaluated our approach to illustrate the benefits of the combination of range and visual features for the task of place recognition. In the next sections the following hypotheses are empirically validated:

Fig. 4. Example images from different sequences, where each row comprises images from one scene. In addition to four visual example images, the range image which corresponds to the rightmost visual image is shown. The scene categories in our setting are (listed from top to bottom) *Corridor, Elevator Area, Entrance Area, PhD Lab, Kitchen, Robot Lab,* and *Student Lab.*

1. Incorporation of range features improves the recognition performance.
2. The Randomized Decision Forest classifier and the SVM classifier achieve the best recognition rates with a combination of different feature types.
3. The use of temporal context information by means of hidden Markov models leads to an important gain in performance.

Our empirical evaluation is based on a place recognition scenario with seven different rooms (classes). The final dataset consists of eight sequences, where

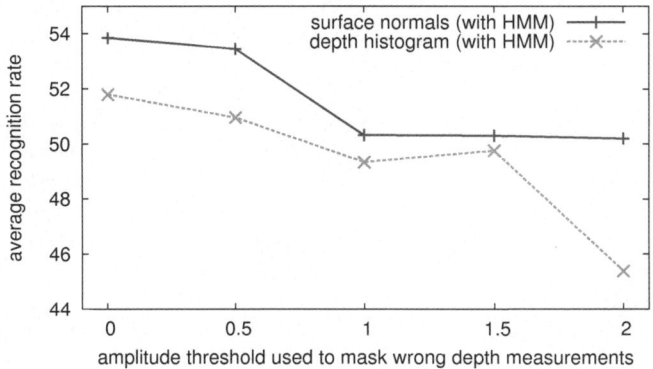

Fig. 5. Influence of additional preprocessing of the ToF data: all depth measurements below a given amplitude threshold are discarded in the computation of range features. A zero threshold corresponds to raw ToF data.

each sequence was captured by navigating a mobile robot through a subset of the rooms. Roughly each second, a PMD[vision] 19k camera and a standard CCD camera obtained range and visual images (Fig. 1). As can be seen in Fig. 4, visual and range images do not contain exactly the same image sections, which is due to the different angle of view of the cameras. Note that a calibration of the cameras was not necessary, because features are calculated from the different sensor images independently.

Training is done on two chosen sequences, which together cover all classes of the dataset. The remaining six sequences were then used to test the recognition performance. To measure recognition performance, unbiased average recognition rate was computed. Since more than one scene is used for testing, the mean of all average recognition rates (one for each sequence) is used to evaluate our system.

5.1   Evaluation of Preprocessing Techniques

Due to the severe noise of the ToF range data, one often has to mask outliers using the amplitude image. All depth measurements with corresponding amplitude value below a predefined threshold are discarded. Nevertheless, we do not apply this preprocessing technique prior to feature computation because it would decrease the recognition performance in our setting.

We analyze this surprising effect in the following experiment. The recognition performance is evaluated for the surface normal feature and the range histogram feature using the RDF classifier with several values of the amplitude threshold. A threshold of zero corresponds to raw data without preprocessing.

The results are illustrated in Fig. 5. and show that the recognition rate decreases if we discard more and more measurements, even erroneous ones. Our place recognition system, therefore, seems to benefit also from wrong measurements which are possible cues of black or critical surfaces.

Table 1. Evaluation of different features types (incl. computation time) with the best classifier result and HMM integration. Features computed on the range image of the ToF sensor are tagged with a preceding $r-$.

Feature type	Avg. Recognition Rate	Time (in sec)
$r-$hist	51.8	0.024
$r-$power	48.5	0.031
$r-$gabor	45.5	0.140
$r-$surface	**53.8**	0.303
power	55.4	0.040
gabor	**64.6**	0.512
feature combination	**67.0**	0.839

Table 2. Table of feature type combinations (among the tested subsets), which lead to the best recognition performances with HMM integration

classifier	Gabor	power	$r-$Gabor	$r-$power	$r-$hist	$r-$surface	result
Parzen	×	×		×			65.4
MLP	×		×	×			65.5
RDF	×				×	×	**67.0**
SVM	×	×			×	×	65.6

5.2   Evaluation of Feature Types and Combinations

In order to evaluate the effects of combined features, we first analyzed the classification performance on each feature type separately. The recognition results are illustrated in detail in Fig. 6 and summarized in Table 1, where only the best (out of four) classifier result is shown. Regarding the range features, our experiments show that the surface normal histogram ($B_\varphi = B_\theta = 10$, $P = 3$) achieves the best place recognition result. However, Gabor and power spectrum features computed using the data from the visual sensor yield a higher recognition performance.

As can be seen in Table 1, feature combination leads to a substantial performance gain over single feature types. The best combination scheme achieved is a recognition rate of 67.0%.

5.3   Evaluation of Different Classifiers

In the preceding section we showed that the combination of different feature types can improve the classification performance. However, the amount of performance gain depends on the used classifier. We also observed that the classifiers achieved best results when only a subset of all feature types were used. By analyzing either a manually chosen list of feature combinations (for RDF and SVM) or by applying a greedy search algorithm on the space of combinations (for Parzen and MLP), we obtained the results shown in Fig. 7 with corresponding

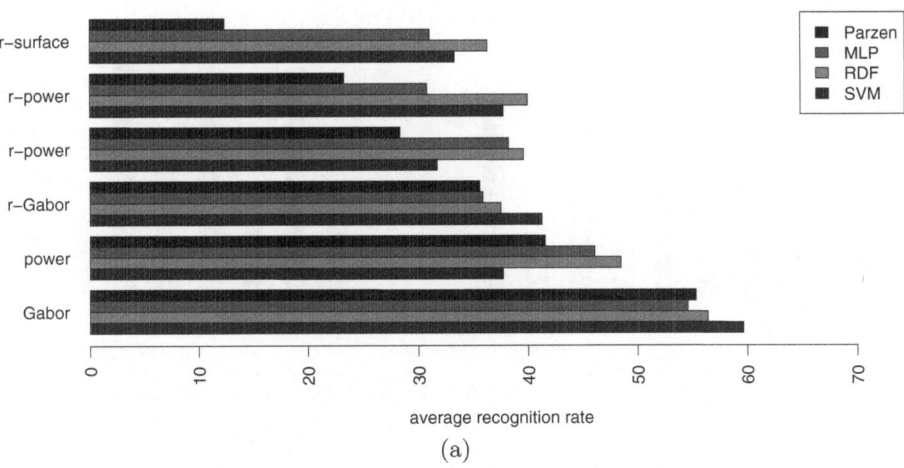

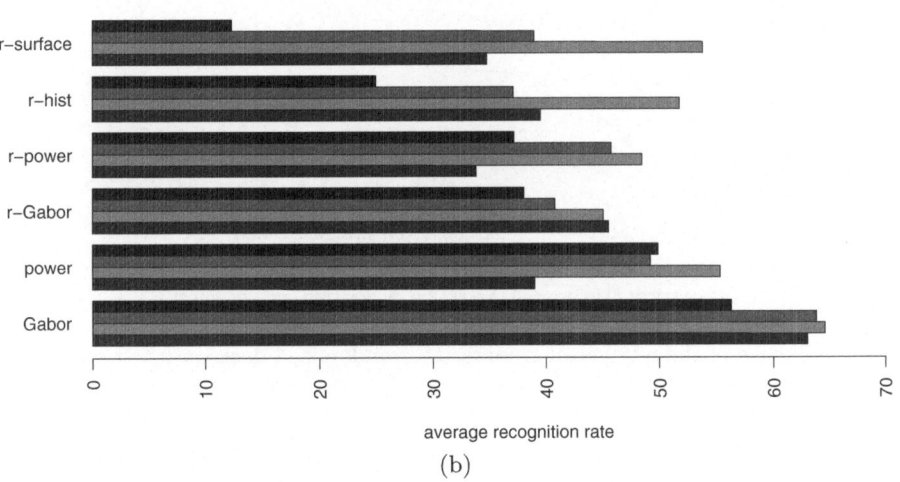

Fig. 6. Performances of single features types without hidden Markov model (a) and with hidden Markov model (b)

combinations listed in Table 2. These average recognition rates suggest that the RDF is the appropriate classifier for our scene recognition task.

In order to further evaluate the power of range information, we removed all range features from the used feature type subsets mentioned above, i.e. only a combination of visual features remains. The average recognition rates in Fig. 7 (visual) illustrates a drop in classification performance for all classifiers except SVM. These results clearly show the advantage of our multi-sensor approach.

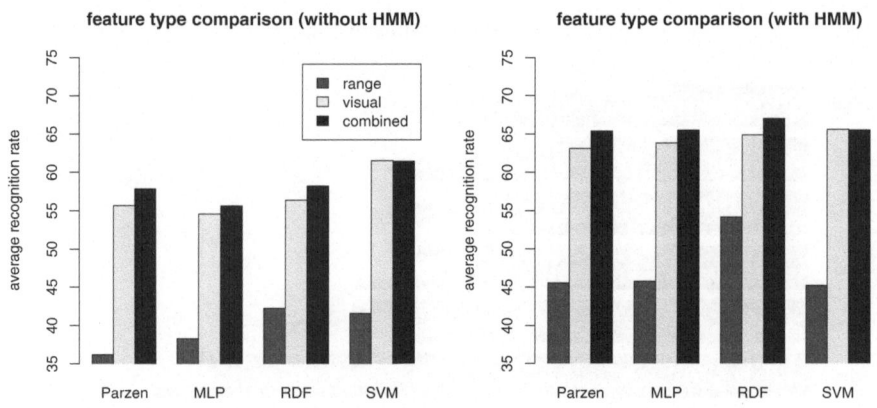

Fig. 7. Comparison of various feature type combinations (sensor-specific and mixed)

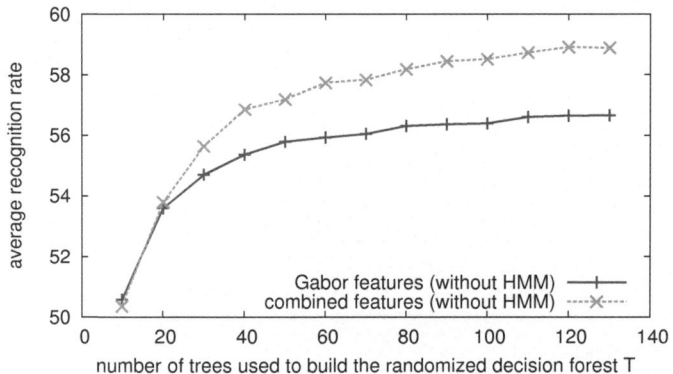

Fig. 8. Influence of the number of trees for the randomized decision forest with visual Gabor features or a combination of range and visual features

It can be also seen that without the integration of the hidden Markov model the recognition performance decreases substantially. This observation highlights the importance of temporally contextual information in our scene recognition task.

Finally, in order to allow a more detailed analysis of the scene recognition result obtained by the best feature type combination, we computed the confusion matrix for this setting (averaged over 30 results). As can be seen in Fig. 9, the recognition rates for six out of eight rooms vary between 76.9% and 85.3%. The significantly lower overall recognition rate (67.0%) is thus directly related to the low recognition rates of the remaining two categories *PhD Lab* and *Robot Lab*, which tend to be recognized as *Student Lab*. This behavior stems from the

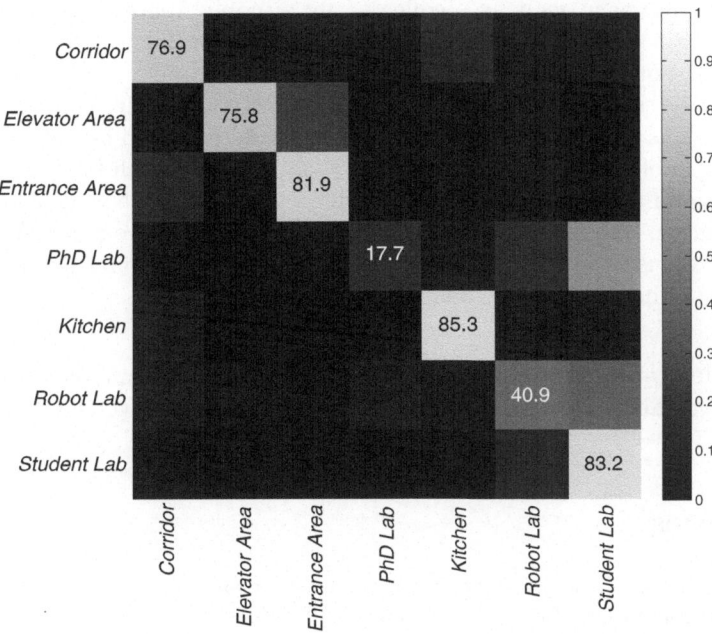

Fig. 9. Confusion matrix for the best feature combination setting (RDF+HMM)

close holistic similarity of these rooms and suggests that more locally receptive features could be promising in order to differentiate between these similar rooms.

5.4   Influence of the Number of Trees

In our previous experiments we used $T = 100$ trees for the randomized decision forest. To investigate the influence of this parameter we perform tests with Gabor and combined features without HMM. The results can be seen in Fig. 8. To cope with the randomization, we average the results of 200 runs for each data point. As can be seen, the generalization performance increases with the number of trees even beyond $T = 100$. However, this effect levels out after a specific size of the forest.

6   Conclusion and Further Work

We presented an approach to place and scene recognition which combines information from both a ToF sensor and a standard visual sensor without calibration. We utilized state-of-the-art feature representations from the field of scene recognition [9,4] and developed a novel description of the range image using planar patches. To show the applicability of our method, we performed experiments with multiple image sequences collected by a mobile robot. The resulting performance gain of the combined feature representation highlights the usefulness of a ToF sensor for the task of place recognition.

As an interesting direction for future research, our feature description of the range image as a histogram of surface normals could be used in conjunction with the principle of spatial pyramid matching [5]. This approach has been shown to lead to a significant performance gain by incorporating rough spatial information within images. The most interesting application of our place recognition system would be to use the probabilities of places as prior information in an object detection setting as proposed in [11].

Acknowledgements. We would like to thank all four anonymous reviewers for their valuable comments, which really helped to improve the quality of the paper.

References

1. Lange, R.: 3D Time-of-Flight Distance Measurement with Custom Solid-State Image Sensors in CMOS/CCD-Technology. PhD thesis, University of Siegen (2000)
2. Prusak, A., Melnychuk, O., Roth, H., Schiller, I., Koch, R.: Pose estimation and map building with a time-of-flight camera for robot navigation. Int. J. Intell. Syst. Technol. Appl. 5, 355–364 (2008)
3. Hegazy, D., Denzler, J.: Generic 3d object recognition from time-of-flight images using boosted combined shape features. In: Proc. of VISAPP, pp. 321–326 (2009)
4. Torralba, A., Murphy, K.P., Freeman, W.T., Rubin, M.A.: Context-based vision system for place and object recognition. In: Proc. of ICCV, pp. 273–280 (2003)
5. Lazebnik, S., Schmid, C., Ponce, J.: Beyond bags of features: Spatial pyramid matching for recognizing natural scene categories. In: Proc. of CVPR, pp. 2169–2178 (2006)
6. Murray, D.R.: Patchlets: a method of interpreting correlation stereo three-dimensional data. PhD thesis, The University of British Columbia, Canada (2004)
7. Hetzel, G., Leibe, B., Levi, P., Schiele, B.: 3D object recognition from range images using local feature histograms. In: Proc. of CVPR, vol. 2, pp. 394–399 (2001)
8. Mezrich, J., Carlson, C., Cohen, R.: Image descriptors for displays. Technical Report PRRL-77-CR-7, Office of Naval Research (1977)
9. Oliva, A., Torralba, A.: Modeling the shape of the scene: A holistic representation of the spatial envelope. IJCV 42, 145–175 (2001)
10. Redies, C., Hasenstein, J., Denzler, J.: Fractal-like image statistics in visual art: similarity to natural scenes. Spatial Vision 21, 137–148 (2007)
11. Torralba, A.: Contextual priming for object detection. International Journal of Computer Vision 53, 169–191 (2003)
12. Wainwright, M.J., Schwartz, O., Simoncelli, E.P.: Natural image statistics and divisive normalization: Modeling nonlinearity and adaptation in cortical neurons. In: Rao, R., Olshausen, B., Lewicki, M. (eds.) Probabilistic Models of the Brain: Perception and Neural Function, pp. 203–222. MIT Press, Cambridge (2002)

13. Bishop, C.M.: Pattern Recognition and Machine Learning (Information Science and Statistics), 1st edn., Springer, Heidelberg (2007)
14. Schimek, M.G.: Smoothing and Regression: Approaches, Computation, and Application. Series in Probability and Statistics. Wiley, Chichester (1996)
15. Breiman, L.: Random forests. Machine Learning 45, 5–32 (2001)
16. Geurts, P., Ernst, D., Wehenkel, L.: Extremely randomized trees. Maching Learning 63, 3–42 (2006)
17. Platt, J.: Probabilistic outputs for support vector machines and comparison to regularize likelihood methods. In: Smola, A., Bartlett, P., Schoelkopf, B., Schuurmans, D. (eds.) Advances in Large Margin Classifiers, pp. 61–74 (2000)

Shadow Detection in Dynamic Scenes Using Dense Stereo Information and an Outdoor Illumination Model

Claus B. Madsen[1], Thomas B. Moeslund[1],
Amit Pal[2], and Shankkar Balasubramanian[2]

[1] Computer Vision and Media Technology Lab
Aalborg University, Aalborg, Denmark
cbm@cvmt.aau.dk
www.cvmt.aau.dk/~cbm
[2] Department of Electronics and Communication Engineering
Indian Institute of Technology Guwahati, Assam, India

Abstract. We present a system for detecting shadows in dynamic outdoor scenes. The technique is based on fusing background subtraction operations performed on both color and disparity data, respectively. A simple geometrical analysis results in an ability to classify pixels into foreground, shadow candidate, and background. The shadow candidates are further refined by analyzing displacements in log chromaticity space to find the shadow hue shift with the strongest data support and ruling out other displacements. This makes the shadow detection robust towards false positives from rain, for example. The techniques employed allow for 3Hz operation on commodity hardware using a commercially available dense stereo camera solution.

Keywords: Shadows, stereo, illumination, chromaticity, color.

1 Introduction

Shadows are an inherent part of images. Especially in outdoor vision applications shadows can be a source of grave problems for the processing and analysis of video data. For humans, though, shadows represent a significant cue to understanding the geometry of a scene, and to understanding the illumination conditions, which in turn helps processing the visual data. In this paper we present an approach to accurately identifying shadow regions in outdoor, daylight video data in near real-time (presently around 3 Hz, with potential for significant improvement). The main contributions of this work lie in utilizing a combination of color and dense depth data from a stereo rig for an initial, rough shadow detection, combined with a model-based chromaticity analysis for the final, precise shadow pixel identification.

Work on detection of shadows can be divided into techniques for detecting dynamic shadows (cast by moving objects) and static shadows (cast by static

R. Koch and A. Kolb (Eds.): Dyn3D 2009, LNCS 5742, pp. 110–125, 2009.

scene objects such as buildings). Static shadow detection work is based on single images and is typically more sophisticated in the use of physically based illumination/reflection. It is also typically computationally heavy techniques. Dynamic shadow detection work is naturally based on image sequences and utilizes somewhat simplistic illumination models which at best correspond poorly to real conditions, especially for outdoor scenery. These techniques all employ background subtraction based on a trained background model, a concept which is problematic for very long outdoor image sequences due to drastic illumination changes, precipitation, foliage changes, etc.

The ideas proposed in this paper can be operated in two modes: one based on background subtraction with a trained model, and one based on image differencing with no training. We shall focus on the former mode in the presentation and return to the latter mode in section 5.

Detecting static shadows is in principle difficult as it is theoretically impossible to definitively determine whether a region in an image is a bright surface in shadow or a dark surface in direct light. Regardless, promising results on single image shadow detection and removal has been presented over the recent years. A single image shadow removal technique is presented in [3] but requires a very high quality, chromatically calibrated camera, and does not handle soft shadows (penumbra regions). The technique presented in [11] distinguishes between cast shadows (on a plane) and self-shadowing, but is tested on somewhat simple scenarios, and it too does not handle soft shadows. Not being able to handle soft shadows is a severe problem for outdoor scenes in partly overcast conditions.

Single image shadow detection in scenes with soft shadows is addressed in [9,8], demonstrating successful shadow detection (and removal) on single images of non-trivial scenes. Unfortunately, the approach requires manual identification of training areas in the image (areas where the same material is visible in shadow as well as in direct sunlight).

So, the state-of-the-art in single image shadow work is that it does not really handle soft shadows, or requires manual training. Our method handles soft shadows very well, and we demonstrate the even quite subtle shadows in allmost overcast conditions can be detected. Furthermore we demonstrate that our method, in the no-background-model mode mentioned above, can generate the necessary input for the technique described in [8] thus eliminating the need for manual boot-strapping.

Dynamic shadow detection based on image sequences has recently received much attention especially in the surveillance literature. Here there is a need for detecting and tracking objects in a scene and one of the key problems has turned out to be false positives due to shadows [10,7]. Many of the approaches suggested for shadow segmentation are based on the idea that a pixel in shadow has the same color as when not in shadow, but at a lower intensity [10,6,4]. Such an illumination model is very simplistic (assumes all light sources to be white). This is a severe assumption, which is totally violated in outdoor scenes, and tests presented in in these works are also either on indoor scenes or on outdoor scenes in overcast conditions, where the assumption roughly holds.

In non-overcast outdoor scenes regions in shadow (blocked from direct sunlight) exhibit a blue hue shift due to the differences in the spectrum of the light coming from the sky, and the light coming from the sun. This fact is incorporated into the work presented in [5], which utilizes the blue shift for outdoor scenes for separating foreground from shadow using using background subtraction.

All previous work on dynamic shadow detection utilized a trained background model in some form. By combining depth information with color information the techniques presented in this paper makes it much simpler to distinguish between foreground and shadow, allowing us to operate with a much less well-trained background model (and thus more robust towards illumination changes, precipitation, etc.). And, as mentioned, we can even detect a substantial part of the shadows without any kind of background model.

The inspiration for the work presented in this paper came from two sides: 1) in dynamic outdoor scenes (time sequence video of scenes with moving objects) some shadows will move, which provides a unique opportunity to study the same pixel under both shadow *and* non-shadow conditions (making it possible to estimate the shadow hue shift automatically), and 2) when combining color information with dense depth information shadow candidate regions can be identified from the observation that a cast shadow represents a change in color channel values but not in depth (for shadows that fall on static surfaces in the scene).

The paper is organized as follows. In section 2 we present the setup used and give a brief introduction to the methods presented later in the paper. Section 4 describes how the color and depth information is combined to detect shadow candidate regions, and how chromaticity analysis is used to finally identify shadow pixels. We then present and discuss some results in section 5, followed by conclusions.

2   Setup and Overview of Approach

The setup for this work is centered around a commercial stereo rig from Point Grey Research INC., [12], see figure 1. The Bumblebee XB3 real-time dense stereo camera delivers rectified stereo image pairs via FireWire at up to 16 frames per second, depending on resolution. In this work we operate with a 640x480 resolution, resulting in a stereo frame rate of approx. 10 Hz. Using the accompanying SDK for the stereo camera disparities can be computed at a per pixel level using correlation techniques. On an Intel Core Duo 2 2.3 GHz machine running Windows XP SP2, equipped with 2 GByte RAM, the disparities are computed in around 50 milliseconds, so the limiting factor is the 10 Hz transfer of rectified stereo images from the camera. All RGB and disparity images shown in this paper represent the view of the right camera of the stereo rig. The disparity values employed in this work are in 16 bit resolution (subpixel disparity information) as the Bumblebee XB3 SDK offers this functionality.

It is assumed that the camera is static relative to the scene. It is also assumed that the scene contains a substantial amount of static surfaces (objects that do

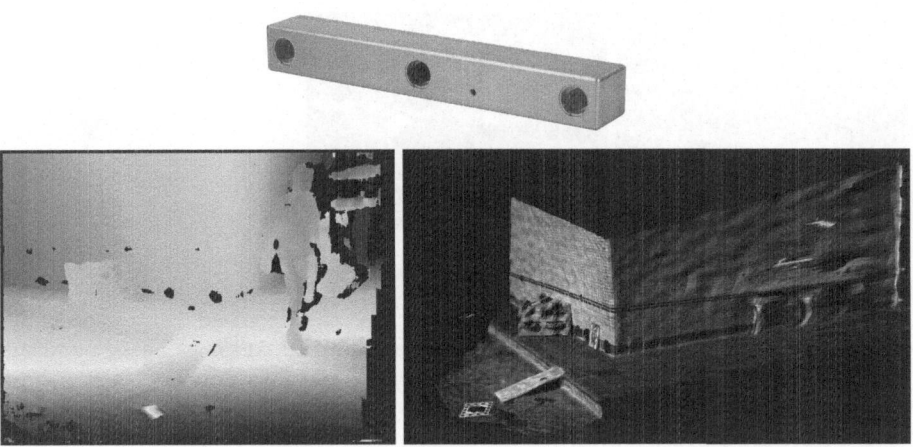

Fig. 1. Top: The commercially available Bumblebee XB3 stereo rig from Point Grey Research, Inc. at approximately 3000 USD. Bottom left: Pseudo-colored disparity image of an outdoor scene. The dark red patches represent undetected disparities due to pixels being over-exposed or under-exposed, lack of texture, or pixels representing scene points that are not visible in both cameras (occlusion). Bottom right: 3D mesh constructed from disparity image and textured with the color information from the RGB image. This scene is a frame with no person.

not move), and that dynamic objects are also present in the scene. In general the scene will contain shadows cast by static objects, as well as shadows cast by dynamic objects. The techniques presented in this paper are able to detect shadows cast by the dynamic objects, although section 5 demonstrates how the generated results can be used to detect the static shadows as well.

The approach presented here has 4 main steps. First we employ a background subtraction method on the color information (the RGB image). Next we perform a similar step on the dense per pixel disparity information, where the background image in this case represents an acquired depth model of the scene (disparities are proportional to metric depth, so there is no reason to spend computational resources on scaling disparities to metric depth unless metric information is required for some other processing step unrelated to the core shadow detection). The third step is to combine the results of the two background subtractions which allows us to interpret the nature of each pixel: foreground, background, shadow. This is illustrated in figure 2. The fourth and final step is to evaluate some chromaticity (normalized color channel information) characteristics of the segmented shadow pixel population to eliminate those pixels that do not conform to an illumination model which predicts the overall behaviour of regions as they transition from being exposed to direct sunlight to being in shadow.

Subsequently, the different steps are elaborated in further detail. While we have an operational C++ real-time implementation of the described approach some of the results shown in this paper are generated by a similar Matlab

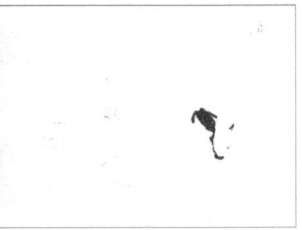

Fig. 2. Processing results from frame 137 out of a recorded sequence of 200 frames. Left: The current frame where the person has entered the scene and has splashed water from the bucket onto several surfaces in the scene. Middle: result from background subtraction, segmented into foreground object (grey) and shadow candidate pixels (white), which include all the detected water splashes since the water has caused the surfaces to change appearance. Left: pixels identified as being in shadow after chromaticity analysis. Note that the shadow cast by the camera tripod is a static shadow, and as such is not detected by the presented approach.

implementation from which we have easier access to specific intermediate results and can generate illustrative visuals for the paper.

3   Theoretical Framework and Fundamental Assumptions

The work presented in this paper rests on a fundamental radiometric model of the radiance of points in a scene illuminated by a combination of sunlight and sky light. This model, together with some assumptions that are made, are described in this section.

It is assumed that the images represent an outdoor scene subjected to daylight illumination. It is also necessary to assume that the materials represented in the scene are predominantly diffuse (exhibit Lambertian reflectance distribution characteristics). We do *not* require the albedos (diffuse reflectances) of the surfaces in the scene to be constant over time. In fact we clearly demonstrate that our approach can avoid errornously detecting/hallucinating shadows in areas where the surface has simply changed appearance from suddenly becoming wet (as they would in rainy conditions).

Concerning the images it is furthermore assumed that they are properly exposed, i.e., important areas in the image are allowed to be neither severely overexposed (color channel values near 255) nor severely under-exposed (values near 0). We will also assume that it is a fair assumption that the camera is linear, in the sense that there is a linear relationship between the radiance of a surface and the pixel value assigned to the image point of that surface.

For a linear camera the pixel value in some color channel is proportional to the reflected radiance of the surface being imaged, and radiance is measured in $W/(m^2 \cdot sr)$. In a setting as described above it is possible to formulate the value, P_r, of a pixel as follows, using subscript r to indicate elements particular to the red channel (green and blue channel being similar):

$$P_r = c_r \cdot \rho_r \cdot E_r \cdot \frac{1}{\pi} \qquad (1)$$

where ρ_r is the diffuse albedo of the surface point being imaged (ratio of outgoing radiosity to incoming irradiance), and E_r is the incoming irradiance in the red channel. Thus, $\rho_r \cdot E_r$ is the reflected radiosity. Dividing this by π $[sr]$ yields the reflected radiance of the surface (since the radiosity from a diffuse surface is π times the radiance of the surface). Finally, c_r is the (typically unknown) scaling factor translating the measured radiance into pixel value (0 to 255 range for an 8 bit camera) for a linear camera. This scaling value depends on the aperture of the lens, the shutter speed, the gain, the white-balancing etc. of the camera.

In the kind of outdoor daylight setting we are addressing in this paper the total incoming irradiance at a point is a sum of two contributions, $E_r = E_r^{\mathrm{sun}} + E_r^{\mathrm{sky}}$, again using subscript r for red color channel as example. The amount of irradiance received from the sun, E_r^{sun}, depends on several factors: the radiance of the sun, how large a fraction of the sun's disk is visible from the point in interest (if the sun's disk is completely occluded the point is in full shadow, also called umbra), and on the angle between the surface normal at the point and the direction vector to the sun from the point. If the sun's disk is only partially occluded the point is in the penumbra (soft shadow).

We shall return to this formulation in section 4.3, where we use it to justify our approach to letting shadow candidate pixels vote for a shadow hue shift which can be used to dismiss pixels that are in fact not in shadow in a particular frame.

4   Methods

As described in section 2 we initially segment each frame into background, foreground, and shadow. This is performed by combining the results from a background subtraction process on both the color image information and on the disparity image information. The two background subtraction processes are described below.

4.1   RGB Background Subtraction

We apply the Codebook method [6] since it has been shown to outperform other background subtraction methods [2]. The method contains three steps: modeling the background, pixel classification and model updating.

Each pixel is modeled as a group of codewords which constitutes the codebook for this particular pixel. Each codeword is a cylindrical region in RGB-space and for each new frame each pixel is compared to its codebook. If the current pixel value belongs to one of the codewords it is classified as background, otherwise foreground.

The codebooks are built during training and updated at run-time. The training phase is similar to the pixel classification except that a foreground pixel results in the construction of a new codeword and a background pixel is used to

Fig. 3. Left: Pixels (frame 137) determined by the Codebook method as being different from the trained color background model. Right: Pixels (also frame 137) determined as being different from the disparity background model.

modify the codeword it belongs to using a standard temporal weighting scheme. The codebooks generated in this way during training will typically fall into three categories:

Static codebook. For example a pixel representing a road with no shadows or occlusions. Typically only one codeword is used.

Quasi-static codebook. For example a pixel containing the sky, but sometimes occluded by vegetation due to wind gusts. During training typically two codewords will be constructed for this codebook, one for the sky and one for the vegetation.

Noisy codebook. One of the above combined with noise in the form of a pedestrian, car etc. passing by the pixel or noise due to incorrect segmentation. The result will be an often high number of codewords for this codebook.

To handle the noisy codebooks a temporal filter is applied. It is based on the Max Negative Run-Length (MNRL), which is the longest time interval in which a codeword has not been activated. The filter effectively removes codewords with little support during the training phase, such as passing pedestrians.

Normally it is difficult to tune the sensitivity of the Codebook method (and other background subtraction methods for that matter) particularly due to problems with shadows. In this work, however, this is less of a problem since over-segmentation is actually a desired effect. We therefore tune the method to detect even small changes, and we do not train with shadow regions. This effectively results in a segmentation of the dynamic foreground object including its shadow, see figure 3 for an example.

4.2   Disparity Background Subtraction

We also apply the Codebook method for depth-based background subtraction. Here we only apply one codeword per pixel and its value is not the actual depth

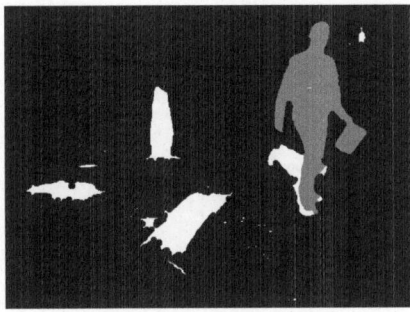

Fig. 4. Classification of pixels into foreground (grey), background (black) and shadow candidates (white)

value, but rather the disparity. The disparity background model is learned as the Median of a number of training images.

Normally a disparity map contains undefined pixels, due to e.g., noise. We therefore smooth the disparity map to obtain more consistent data, and the pixels classified as different from the background also need some clean-up using standard morphological operations. We have employed erode followed by dilate with a radius 5 pixels disk structuring element.

4.3   Shadow Classification

For each pixel in an input image we now have two TRUE/FALSE values with respect to whether the pixel is different from the RGB background model and whether it is different from the disparity background model. From this we can infer the pixel's type (foreground, background, shadow) as shown in table 1. In figure 4 a labeling based on table 1 is shown. The rational behind this table can be formulated as follows: if the color of a pixel has changed, but there is no change in disparity, then the pixel has gone from direct light to shadow. If there is a change in disparity but not in color it can be argued whether to classify it is foreground or background. We have chosen background, since disparity data is less robust than color data, at any rate if there is not change in color it cannot represent a shadow.

Table 1. Classification scheme based on results of background subtraction on color and depth data. It should be noted that only color changes that represent *decreased* intensity are valid shadow candidates.

		Change in RGB?	
		Yes	No
Change	Yes	Foreground	Background
in depth?	No	Shadow	Background

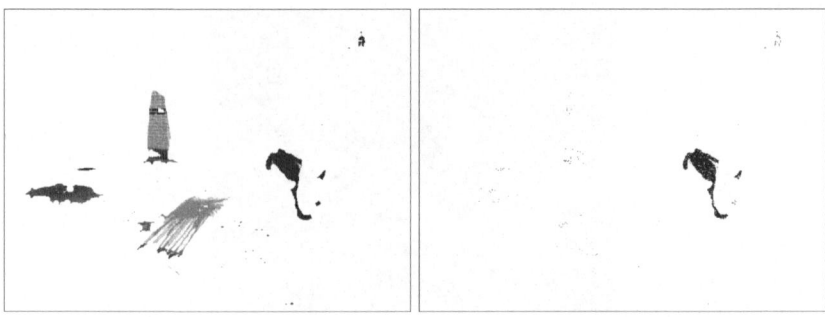

Fig. 5. Left: population of shadow pixel candidates after logical classification based on background subtraction on RGB and disparity data. Right: population of shadow pixels after analysis of the permissible shift in the log chromaticity plane.

After this logical classification we are left with a population of dynamic (cast by a dynamic object) shadow candidate pixels for each frame. These pixels are shadow *candidates* only. Predominantly two things can cause pixels to falsely be labeled as shadow pixels: 1) the albedo of the surface changed (for example due to the surface becoming wet), or 2) imperfections of the disparity data causes the disparity background subtraction to produce sub-optimal results (see for example figure 3). All pixels labeled as shadow candidates are shown with their RGB values in figure 5.

We address the problem of rejecting the non-shadow pixels by returning to the formulation of the value of a pixel as given in section 3. In log chromaticity space the pixel values become:

$$
\begin{aligned}
r &= \log(P_r/P_g) \\
&= \log(P_r) - \log(P_g) \\
&= \log(c_r) - \log(c_g) + \log(\rho_r) - \log(\rho_g) + \log(E_r) - \log(E_g) \quad (2) \\
b &= \log(P_b/P_g) \\
&= \log(c_b) - \log(c_g) + \log(\rho_b) - \log(\rho_g) + \log(E_b) - \log(E_g) \quad (3)
\end{aligned}
$$

The next observation is that the background image depicts the scene free of dynamic shadows. Thus, if for a given frame a pixel has been classified as a shadow candidate by applying the rules in table 1, we have the same pixel in two different versions: 1) a version in direct sun light from the background image, where $E = E^{\text{sky}} + E^{\text{sun}}$, and 2) a version in shadow from the current frame, where $E = E^{\text{sky}}$ (with appropriate indexes for respective color channels). If we subtract the chromaticity values of these two versions for a given pixel:

$$
r^{\text{sky}} - r^{\text{sun + sky}} = \log\left(\frac{E_r^{\text{sky}}}{E_r^{\text{sky + sun}}}\right) - \log\left(\frac{E_g^{\text{sky}}}{E_g^{\text{sky + sun}}}\right) \quad (4)
$$

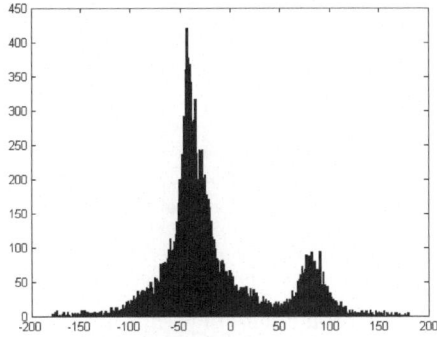

Fig. 6. Histogram over chromaticity displacement vector orientations, measured in the range from -180 degrees to +180 degrees. In this case the histogram has two peaks, one at around -45 degrees, corresponding to the darkening of the wooden plate as a result of a water splash, and one at around 90 degrees, corresponding to the actual blue shift of the shadow.

$$b^{\text{sky}} - b^{\text{sun}} + \text{sky} = \log\left(\frac{E_b^{\text{sky}}}{E_b^{\text{sky} + \text{sun}}}\right) - \log\left(\frac{E_g^{\text{sky}}}{E_g^{\text{sky} + \text{sun}}}\right) \qquad (5)$$

Looking at eqs. 4 and 5 we see that by subtracting the log chromaticity coordinates of the two different versions of a pixel we get a 2D vector in log chromaticity space which is independent of camera properties (the $c_{r/g/b}$ scaling constants), and independent of the material properties (the $\rho_{r/g/b}$ albedos). The only things that influence these log chromaticity displacement/hue shift vectors are the irradiances. Furthermore the depth of the shadow (determined by the amount of sun disk occlusion and the angle between the surface normal and the direction vector to the sun) only influences the length of this displacement vector, not the direction.

With these observations we compute the orientations of all these displacement vectors (one for each of the shadow pixel candidates) and form a histogram of the orientations in the range from -180 to +180 degrees, see figure 6. The number of bins in the histogram is set to the number of shadow candidate pixels divided by 50 to ensure a reasonable number of candidates in each orientation bin. The minimum number of bins is 10, though, to handle the case of very few detected shadow candidates.

Since the spectrum of light from the sky is dominated by wavelengths in the blue channel pixels that go from direct sun light to shadow conditions will undergo a blue shift, which in terms of the histogram in figure 6 corresponds to displacement orientation near +90 degrees (in rb chromaticity space blue chromaticity is upwards). We therefore search the histogram to find a local maximum near 90 degrees. The chosen peak corresponds to the chromaticity shift pixels undergo when they transition from shadow to direct light. All pixels whose displacement vectors are not close to (in our system within 20 degrees of) this

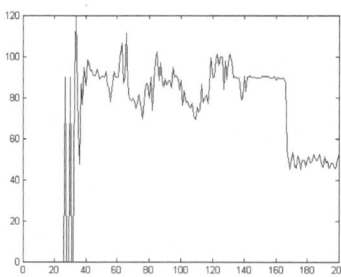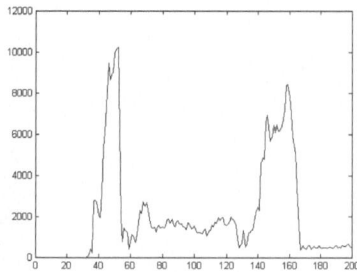

Fig. 7. Left: The detected log chromaticity blue shift direction (in degrees) as a function of frame number. Right: The number of verified shadow pixels in the scene as a function of frame number. When there are too few shadow pixels (less than around 500) the direction cannot be detected robustly.

"most voted for direction" are classified as not shadow pixels and are removed from the shadow pixel population. The remaining pixels all exhibit the same behaviour in log chromaticity space and are thus all consistent with the illumination model. In figure 5 it can be seen how this chromaticity analysis removes wrongly detected shadow pixels, especially those corresponding to all the water splashes (which have changed the albedo of the surfaces).

5   Results and Discussion

This section will address a set of relevant issues in relation to the presented techniques. We take a closer look at the estimated blue shift direction over a time sequence, we address overlapping shadows, then discuss long time sequences and demonstrate how this work can be operated in a mode with no background model, and finally we discuss some of the assumptions made in this work.

5.1   Blue Shift Direction

The proposed automatic approach to finding the blue shift direction is remarkably robust. Figure 7 plots the chosen direction for the 200 frame sequence used in the above description. The same figure also shows the development in the number of verified shadow pixels per frame through the sequence, and it is essentially seen that the blue shift direction is found robustly whenever there is a sufficient number of shadow pixels available in the scene.

5.2   No Background Model Mode

Our approach as described this far is based on background subtraction in both color and disparity data. As a result shadows that are static, i.e., part of the background image model, do not get detected. Another drawback of using background subtraction is that for very long image sequences (days, weeks, months ...) it can

Fig. 8. In partly overcast, windy conditions illumination changes can be drastic. These two frames are 15 frames apart, corresponding to a time difference of 1.5 second. On average the latter image is 50% brighter than the former.

be difficult to maintain the background model due to highly varying illumination, precipitation, seasonal changes, etc. Figure 8 shows how drastically the illumination can change from one to second to the next, making classical background subtraction very difficult, if not impossible.

To address this problem we demonstrate that image differencing can be employed instead of the background subtraction step. The sequences used in this paper are recorded with 10 frames per second. If we perform image differencing on color and disparity images by subtracting frame T (current frame) and frame $T - \Delta T$ (some frames old), and perform everything else similar to what has been described in the paper, we can detect dynamic shadows with no training of background models at all. Δ can be adjusted to find a compromise between being very robust towards rapidly changing illumination (small ΔT) and detecting all of the shadow area (large ΔT). If ΔT is small compared to the movements of shadows in the scene the shadows in the two frames will overlap, and only part of the shadow will be detected (the part which is in shadow in frame T but not in frame $T - \Delta T$). In this paper we have used a ΔT of 0.5 second, i.e., we perform image differencing with a 5 frame delay. Figure 9 shows some detection results from a sequence acquired under highly varying illumination conditions.

5.3   Detecting All Shadows

By definition our approach only detects the dynamic shadows, regardless of using the background model or the image differencing mode. To address this problem the technique presented here can be combined with an implementation of the technique described in [8]. That method, as described in section 1, requires manual initialization (ratios of sky to sun-plus-sky irradiances for each color channel).

A by-product of the shadow detection technique described here is that it can provide those ratios. These ratios are straight forward to compute, as they are just the per color channel averages of the ratios of detected dynamic shadow

Fig. 9. Dynamic shadow detection based on image differencing (frames 180, 520, and 1741).

pixel values to their non-shadow pixel values. In the case of using the image differencing mode: if image differencing between frame T and frame $T - \Delta T$ results in a pixel being classified as dynamic shadow, then compute the per color channel ratio of pixel value in frame T to the pixel value in frame $T - \Delta T$. For a diffuse surface this ratio equals the sky to sun plus sky irradiance ratio, see section 3. The average of these ratios for all dynamic shadow pixels in a given frame provides the initialization information for the graph cuts technique from [8].

Figure 10 shows the results from using the dynamically detected shadows to boot-strap the graph cuts based shadow segmentation and removal technique, which is capable of handling soft shadows.

5.4   Assumptions Revisited

As described in section 3 this work rests on a number of assumption that are worth discussing. The assumption of the camera being static makes it possible to employ background models or to use simple image differencing as shown above. It would be possible to extend this work to a camera placed on a pan-tilt unit. Omni-directional depth background models are employed in e.g., [1]. If the mounted on a pan-tilt unit a spherical representation of the color and depth background model could be composed by scanning in all directions. If using the image differencing mode optical flow techniques could be employed to compute the overlap between the current frame and the delayed frame used for subtraction. This way the dynamic shadows in the overlap region could be detected and the information from the shadow pixels could then be used for detecting all shadows in the current frame as described in section 5.3.

Fig. 10. First row: original images from three different sequences. Second row: shadows detected by approach described in this paper. Third row: shadows removed with graph cuts based approach. Fourth row: level of shadow (brighter areas represent a deeper shadow level).

A fundamental assumption for this work is that the surfaces are Lambertian (diffuse). We have demonstrated on a number of real outdoor sequences that our model works on a large range of naturally occurring materials in outdoor scenes. Material such as concrete and grass are far from Lambertian when viewed close-up, but at a certain distance they overall display diffuse reflection behaviour because of the surface roughness. Glass and metal surfaces pose a real problem, but the stereo camera can typically not produce valid disparity information from such surfaces and the risk of falsely detecting shadows on such materials is not high (we do not allow a pixel to be classified as shadow if there is no valid disparity value for it). We are presently working on developing a technique for detecting, over long image sequences, pixels that do not conform to a diffuse reflection assumption (i.e., do not over the entire sequence consistently vote for the same illumination model as the majority of the pixels in the scene).

Finally, the theoretical framework is based on an assumption of the camera having a linear response curve, and that the color channels are independent. This is typically only true for very high quality cameras, and certainly the cameras in the Bumblebee stereo rig are not designed for color vision applications. Regardless, we have demonstrated that the model works quite well even with cameras of such low quality.

6   Conclusion

We have presented a technique for detecting shadows in dynamic scenes. The main contributions lie in the combination of color and disparity data analysis, and in the use of qualitative chromaticity plane descriptors for ruling out false positives in the shadow pixel populations. A powerful feature of the proposed approach is its ability to handle albedo changes in the scene, e.g., its robustness towards falsely labeling pixels as shadow in situations where surfaces in the scene have become wet. Another promising feature of the work is that the techniques employed allow for 3Hz operation on commodity hardware using a commercially available dense stereo camera solution.

We conjecture that by enabling vision systems to estimate information about the illumination conditions in the scene the vision systems can be made more robust. In this paper we have demonstrated that it is possible to estimate powerful information concerning the scene illumination in terms of the illuminant direction which can be utilized to verify dynamic shadows and to detect static ones, as well.

Future work will include combining this work with static shadow detection, working with temporal analysis of the detected shadows and illumination information, and using this illumination estimation for realistic augmentation of synthetic objects into the scenes.

Acknowledgments

This research is funded by the CoSPE (26-04-0171) and the BigBrother (274-07-0264) projects under the Danish Research Agency. This support is gratefully acknowledged.

References

1. Bartczak, B., Schiller, I., Beder, C., Koch, R.: Integration of a time-of-flight camera into a mixed reality system for handling dynamic scenes, moving viewpoints and occlusions in real-time. In: Proceedings of the 3DPVT Workshop, Atlanta, GA, USA (June 2008)

2. Chalidabhongse, T.H., Kim, K., Harwood, D., Davis, L.: A Perturbation Method for Evaluating Background Subtraction Algorithms. In: Joint IEEE International Workshop on Visual Surveillance and Performance Evaluation of Tracking and Surveillance, Nice, France, October 11-12 (2003)
3. Finlayson, G.D., Hordley, S.D., Drew, M.S.: Removing shadows from images. In: Heyden, A., Sparr, G., Nielsen, M., Johansen, P. (eds.) ECCV 2002. LNCS, vol. 2353, pp. 823–836. Springer, Heidelberg (2002)
4. Hu, J.-S., Su, T.-M.: Robust Background Subtraction with Shadow And Highlight Removal for Indoor Surveillance. Journal on Advanced in Signal Processing 2007(1), 108–132 (2007)
5. Huerta, I., Holte, M.B., Moeslund, T.B., Gonzàlez, J.: Detection and removal of chromatic moving shadows in surveillance scenarios. In: Proceedings: IEEE ICCV 2009, Kyoto, Japan (September 2009)
6. Kim, K., Chalidabhongse, T.H., Harwood, D., Davis, L.: Real-time Foreground-Background Segmentation using Codebook Model. Real-time Imaging 11(3), 167–256 (2005)
7. Moeslund, T.B., Hilton, A., Krüger, V.: A Survey of Advances in Vision-Based Human Motion Capture and Analysis. Journal of Computer Vision and Image Understanding 104(2-3) (2006)
8. Nielsen, M., Madsen, C.B.: Graph cut based segmentation of soft shadows for seemless removal and augmentation. In: Proceedings: Scandinavian Conference on Image Analysis, Aalborg, Denmark, June 2007, pp. 918–927 (2007)
9. Nielsen, M., Madsen, C.B.: Segmentation of soft shadows based on a daylight- and penumbra model. In: Gagalowicz, A., Philips, W. (eds.) MIRAGE 2007. LNCS, vol. 4418, pp. 341–352. Springer, Heidelberg (2007)
10. Prati, A., Mikic, I., Trivedi, M.M., Cucchiara, R.: Detecting Moving Shadows: Algorithms and Evaluation. IEEE Transactions on Pattern Analysis and Machine Intelligence 25, 918–923 (2003)
11. Salvador, E., Cavalarro, A., Ebrahimi, T.: Shadow identification and classification using invariant color models. Computer Vision and Image Understanding 95, 238–259 (2004)
12. Point Grey Research: Bumblebee,
http://www.ptgrey.com/products/bumblebee/index.html

MixIn3D: 3D Mixed Reality with ToF-Camera*

Reinhard Koch, Ingo Schiller, Bogumil Bartczak,
Falko Kellner, and Kevin Köser

Institute of Computer Science
Christian-Albrechts-University (CAU)
24098 Kiel, Germany
{rk,ischiller,bartczak,fkellner,koeser}@mip.informatik.uni-kiel.de

Abstract. This work discusses an approach to seamlessly integrate real and virtual scene content by on-the-fly 3D scene modeling and dynamic scene interaction. The key element is a ToF-depth camera, accompanied by color cameras, mounted on a pan-tilt head. The system allows to scan the environment for easy 3D reconstruction, and will track and model dynamically moving objects like human actors in 3D. This allows to compute mutual occlusions between real and virtual objects and correct light and shadow generation with mutual light interaction. No dedicated studio is required, as virtually any room can be turned into a virtual studio with this approach. Since the complete process operates in 3D and produces consistent color and depth sequences, this system can be used for full 3D TV production.

1 Introduction

In movie and television productions, there is a great demand to augment a captured scene by including virtual objects and computer generated elements. In movie production, the effects of computer generated augmentation are usually inserted in post production with very high quality. For TV applications, high-end post production is often too expensive, or not feasible at all if the effects are needed on-the-fly during a live broadcast.

For video augmentation, three components are of importance. First, each frame of a video has to be separated into regions showing virtual content and into regions which should retain the real scene. This separation-process is called keying. Furthermore, the tracking of camera motion has to be performed for a proper alignment of virtual and real content. Finally, the interaction of virtual and real content through mutual occlusions, correct shadow casting and reflections is needed for a convincing augmentation. The typical approach to simultaneously solve all of these challenges, is to build a studio environment equipped

---

* This work was partially supported by the German Research Foundation (DFG), KO-2044/3-2 and the Project 3D4YOU, Grant 215075 of the ICT (Information and Communication Technologies) Work Programme of the EU's 7th Framework program.

R. Koch and A. Kolb (Eds.): Dyn3D 2009, LNCS 5742, pp. 126–141, 2009.

with controlled lighting conditions, chroma keying installations, multiple cameras and sophisticated camera tracking systems using markers and sensors. The construction, maintenance and operation of such studios requires a lot of experience, is expensive and, in certain circumstances, impractical. Even if such a studio is available, the issue of mutual interactions between real and virtual content is unsolved unless a complete 3D scene representation can be computed on-the-fly.

In this work we therefore propose an approach exploiting the capabilities of Time-of-Flight depth cameras (ToF-Cameras) [1,2][1]. These devices are capable of providing instantaneous depth maps over a limited field of view (app. $40-50°$) at high frame rates (up to 25 fps). Using such a depth camera in combination with a standard or high-definition video camera, our approach is able to provide all the necessary information to obtain fully automatic 3D object keying and camera tracking, which allows for real-virtual interaction without the need of chroma-keying installation, expensive tracking systems or multiple camera sets. An additional benefit is that each video frame is supplied with full depth information, giving the system the potential to be applied for 3D television production.

MixIn3D addresses all of these challenges. In the next section we will present the system's architecture and the building blocks. Section 3 details how interaction between real and virtual objects can be performed. The presented results are discussed in the concluding section 4.

2   System Architecture

The key components of an augmented reality system are keying, camera tracking and interaction between real and virtual content [3]. Keying is the process in which the foreground object regions are separated from the background in an image. One popular way to achieve this is the chroma keying technique. Here the foreground object is captured in front of a screen of constant color, typically green or blue. Under the assumption that the background color is known, deviation from this assumption can be exploited to detect the image regions occupied by foreground objects and to extract an alpha matte [4]. In cases where the screen's color itself does not fulfill the constant color assumptions, due to lighting conditions, this extraction process can deliver faulty results. To deal with this problem, well lit studio setups (virtual studios) can be used. However these lighting conditions, also captured in the image of the keyed foreground, are difficult to match with arbitrary virtual surroundings. The BBC therefore developed True-matting[2], a chroma keying approach where the constantly colored screen is replaced by retro-reflective cloth, which efficiently reflects light only into the direction, where it was coming from. This way a camera equipped with a ring of LEDs, emitting colored light of low intensity, can be used to generate

---

[1] Companies producing these devices are represented at: www.3dvsystems.com / www.canesta.com / www.mesa-imaging.ch / www.pmdtec.com

[2] www.bbc.co.uk/rd/projects/virtual/truematte/

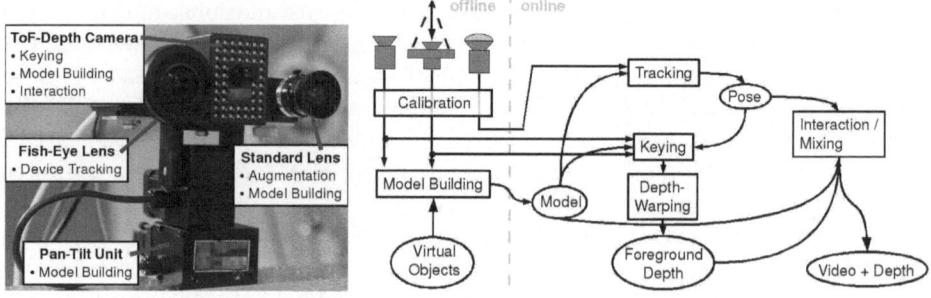

Fig. 1. Major hardware components (left) and their tasks in the system (right)

the required equi-colored background. This technology is expensive and requires to tightly control the environment for a proper segmentation. In order to grant more flexibility, different propositions for the extraction of alpha mattes from images with arbitrary background can be found in the literature [5]. These approaches typically need an initial segmentation of foreground and background and are less reliable. In our approach, we replace the color keying with depth keying, which is discussed in section 2.4.

The 2D keying methods are restricted, since the camera is not allowed to move. To convey a convincing impression of real and virtual content existing together, the camera must be allowed to move, while perspectively correct images of the virtual objects are generated and combined with the real image. This requires the virtual content to be modeled in three dimensions. Furthermore, the real camera pose and projection parameters need to be determined online with the camera motion, so that the virtual object can be rendered perspectively correct. For this purpose, virtual studios are equipped with installations for camera parameter tracking. This tracking equipment ranges from cameras moved by a robot, or using expensive and bulky sensors, to more flexible marker based systems[3] [6]. The installation and proper calibration of such systems is tedious and expensive. Moreover it is required to have a 3D model of the real environment, which contains the relation between the real (physical) scene and the tracking coordinate frame in order to properly align the real and virtual content during the augmentation. In our approach, we automatically model the real environment of the studio by 3D depth scanning. This allows to seamlessly integrate and fuse virtual 3D objects in the real environment for augmentation. Furthermore, the environment model allows for full camera tracking without the need of dedicated markers or robot cameras. Sections 2.3 and 2.5 will address these issues.

The key component of our system is a camera head with a ToF depth camera and a color camera with a field of view of 50°, rigidly coupled together on a computer-controlled pan-tilt unit (PTU). The PTU allows for scanning of the environment in a full sphere of up to 360 × 180°, to overcome the limited field

---

[3] www.orad.co.il

of view of the camera images. In addition, a second camera equipped with a fish-eye lens delivers images with very wide circular field of view of 190°. The use of a fish-eye camera facilitates camera head tracking with high reliability, as noted in [7,8]. Figure 1 on the left shows the camera head with the cameras rigidly mounted to the PTU. On the left and the right side, two HD color CCD cameras, one equipped with a fish-eye lense, the other viewing through a standard lens, are framing the ToF depth camera in the center. The system operates in two modes, as shown in figure 1 on the right. In an offline phase before the actual shooting, the 3D environment model is scanned by combining the depth and color images in a 3D panoramic model. In the online phase, the model is used for depth keying, camera tracking, content mixing and interaction.

2.1   System Calibration

In order to reliably calibrate the intrinsic and relative extrinsic parameters of all cameras of the rig, a procedure based on a known planar checkerboard calibration pattern is applied. We follow the approach discussed in [9] and [10], which was extended to include the calibration of fish-eye lenses. The procedure starts with capturing a sequence of calibration images of the planar checkerboard pattern in different poses and distances. The checkerboard corners are detected and used to find the extrinsic and intrinsic parameters for each camera individually. At this stage the parameter of the ToF-camera and the color camera with the standard lens are computed as in [11,12], while the parameters for the camera with the fish-eye lens are estimated using the results from [13].

These initial parameter estimates are used as the starting-point for a non linear optimization over all parameters, integrating the constraints of fixed relative orientations between all cameras. Furthermore, a final bundle adjustment step optimizes the parameters in an iterative analysis-by-synthesis approach. The initial parameters estimated in one iteration are used to synthesize a color image and a depth image of the known checkerboard pattern. The deviation between the real images and the synthetic data is used to compute an optimized set of parameters for the next iteration. Since the synthesized data is free of noise and every point lying on the checkerboard pattern is contributing to the optimization, the reliability of the calibration is significantly improved.

The depth measurements of the ToF-camera suffers from systematic errors [14], which is not only a constant offset but a higher order function [15]. Therefore the iterative optimization is also estimating the parameters of a spline for depth correction as described in [10]. After calibration, residual reprojection errors of 3D scene objects are well below a pixel, yielding sufficient accuracy for our application.

2.2   Time-of-Flight-Camera (ToF) Principle and Image Fusion

ToF is a sensor principle which delivers dense depth images with up to 25 frames per second and currently a resolution of 176x144 pixel. The camera actively illuminates the scene by sending out incoherently modulated light from an LED-array with a typical modulation frequency of 20 MHz. The light is reflected at the

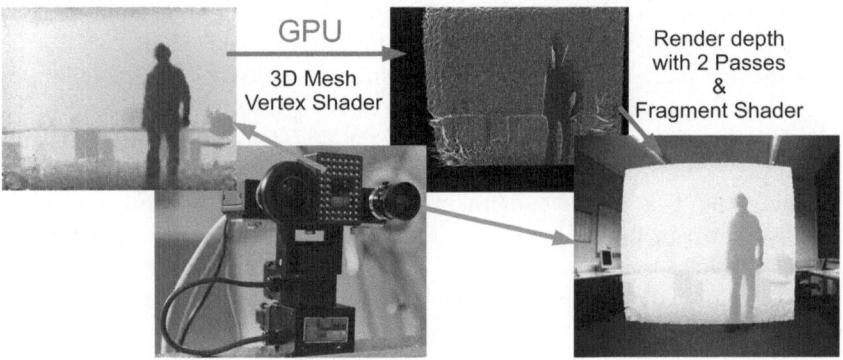

Fig. 2. Depth warping of ToF-image (top left) into the color image (bottom right) with a 3D wire-frame on the GPU (top right)

objects and received by the image sensor of the ToF-camera (cf. [2]). Depending on the object-camera distance, a phase shift in the reflected signal is observable. The ToF-camera is able to extract this phase shift in every pixel and to compute depth from it. For measuring the reflected light, the ToF-camera uses a dedicated semiconductor structure [1].

The phase difference is measured by cross correlation between the sent and received modulated signal of the camera's image sensor. Due to the used modulation frequency, the non-ambiguous range of the ToF-camera is 7.5 meters. From calibration we know that the depth accuracy of the ToF-camera is 20mm or better. Since the resolution of the phase difference measurement is independent from distance, the achievable depth resolution is in a first approximation independent from scene depth. However, since the light intensity fall-off is squared with depth, the signal-noise ratio deteriorates with larger distances and fast sampling rates. Currently we operate the camera with 12.5 fps (80 ms integration time) to keep the depth noise low.

The depth image must be fused with the color images to obtain a combined color-depth video stream. Since the color cameras are displaced from the projection center of the depth camera, a depth-dependent forward-warping is performed that maps the depth pixel into the color camera. We perform the depth warping on the GPU by generating a 3D mesh from the depth camera image, whose z-buffer values are then rendered into the view of the color cameras and scaled for correct depth. This is possible since we obtained all calibration parameters, like projection matrices, radial distortion effects and depth correction from the calibration process with high fidelity. Figure 2 shows an example for depth warping. A challenge is the low resolution of the depth image in comparison to our color camera (1024×768 pixel), where each depth pixel covers an area of about 5×5 color pixels. Thus, the depth image must be upscaled. Depth upscaling to the proper camera viewport is performed automatically during GPU warping, as the wire-frame depth is interpolated during rendering. The low image resolution of depth is not really a problem, as the spatial depth variation is

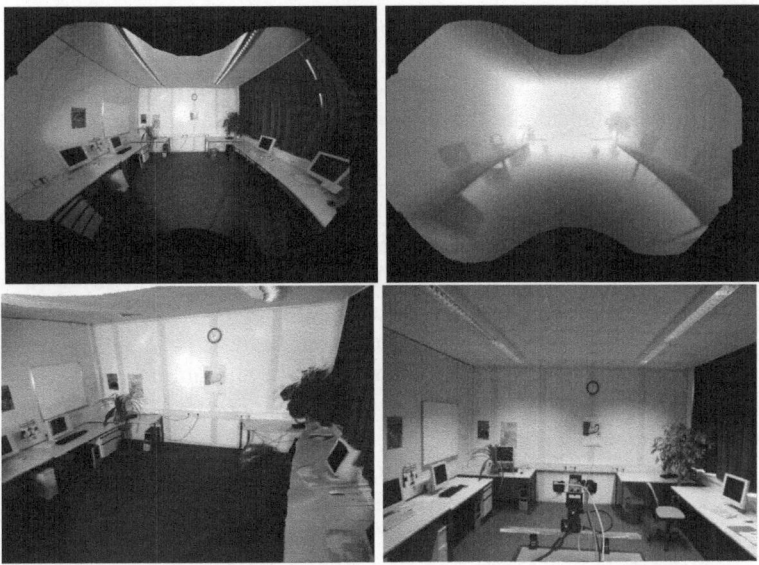

Fig. 3. 3D Panorama of the environment (top row: color and depth) with rendered views of the resulting textured surface (bottom left). Bottom right is showing the position of the camera rig during the model generation.

usually of low frequency w.r.t. color spatial frequencies. The only region where this warping fails are the occlusion regions due to parallax effects of a foreground object occluding parts of the background, as seen in the color camera. However, since in our approach we are only interested in the foreground object area and since we can use the precomputed environment as 3D background model to segment the foreground object, we can eliminate the parallax error. In section 2.4 we will discuss the foreground-background segmentation in more detail.

2.3   Environment Model Building

The 3D environment model is a key feature of our system, which allows for depth keying, camera tracking and 3D interaction. During an offline phase before the actual shooting, a 3D model of the surrounding environment (the studio or any other room) is built by systematically scanning the room with the camera head mounted on the PTU. Each view has only a limited field of view, but stitching together all views will give a complete 3D-color representation of the scene. The PTU yields a very precise orientation stepping of the camera head, and the exact camera pose is obtained by the hand-eye calibration of the camera head. We currently cover an extended field of view of $120 \times 100°$ and allow for image overlap of 50%, which is used for blending of color and depth images. The resulting 3D model is stored as cylindrical panorama for color and depth, since the camera head is not moved during scanning. Note that also a spherical panorama could be used which allows better modeling of floor and ceiling. The panoramic representation limits the operating range to some extent, since some scene parts are occluded, but this does not pose a fundamental limit, and multiple 3D panoramas could be integrated as well. During the online phase, the 3D panorama is unrolled to a full 3D surface representation.

We have chosen this method since it delivers dense and reliable depth very fast, basically at the speed of the image acquisition. The complete scanning takes about 1 minute for the environment. There are, of course, other possibilities to acquire the environment model. Huhle [16] propose to use a rotation sensor scharstein01taxonomy and 3D registration for this. Although very flexible, the danger of errors is high, as the camera motion needs to be computed from the sequence. Stereo estimation, on the other hand (see [17,18]), might fail in untextured regions. A laser scanner, or structured light approaches as presented in [19], might also solve the problem, but will fail in the online phase for dynamic object tracking.

Figure 3 shows an environment panorama and the corresponding surface model. The number of pixels in this panorama can be very high[4], which consequently leads to very large triangle meshes. Therefore a reduction of the redundant triangles is applied [20].

2.4   Actor Segmentation by Depth Keying

During the online phase, a scene is shot with actors moving through the studio. Real-virtual interaction is possible only if we know the 3D position and geometry of the actors in the room, and if we know the surrounding 3D environment. Both is now possible with the proposed system, without any special keying modification like a blue screen. Since the 3D environment was captured before, and the actor is currently observed by the color-depth camera head, depth based segmentation is easily obtained in any environment. The 3D model is rendered into the current camera view and compared with the acquired depth image. Since the fore- and background are separated in depth and since ToF-cameras capture the distance of a surface to the camera regardless of its shape and color and regardless of the lighting conditions, these devices present a good alternative for color keying, even with cluttered backgrounds and changing illumination.

If D_M is a depth map from the background model and D is a corresponding depth map observed by the ToF-Camera, the resulting mask D_{key} is defined as follows:

$$D_{key}(u,v) = \begin{cases} D(u,v)\,, if & D_M(u,v) - D(u,v) > \sigma \\ 0 & ,otherwise \end{cases}.$$ (1)

D_{key} is hereby containing the depth value observed by the ToF-Camera. If the depth difference exceeds some threshold $\sigma > 0$, then the pixel is on the foreground object. This pixel-wise decision is filtered to remove spurious measurement noise.

As discussed in section 2.2, the measurement noise, the low resolution of the depth camera and the parallax error due to warping attribute to keying errors at the object boundary. Thus, the object boundary might contain segmentation

---

[4] In this work we use 4096 × 3072 pixel panoramas.

Fig. 4. Depth keying: Original color and depth image. Bottom: depth-based segmentation.

errors that will be visible in the final rendering. Due to the low resolution, a 5 pixel boundary error might occur in the segmentation which is not acceptable for keying. However, we know that we observe an unknown foreground object superimposed on the known background. Thus, we can compare the current color image with the background image and obtain an improved object boundary. This is however sensitive to illumination changes and shadows eventually caused by the person. As can be seen in figure 4 at the bottom left image the segmentation is not perfect, mainly due to the signal noise. Situations in which foreground and background are connected (e.g. the feet of the person) are difficult to handle and further work is needed. A possible enhancement would be to perform the refinement in a different colorspace (e.g. HSL) which is more invariant towards illumination changes. Alternatively a bilateral filter as described in [21] is used to segment the boundary more precisely. This is very time consuming and not possible in real-time. Additionally depth super-resolution upsampling like [22] can help for better segmentation.

Once the object is segmented, a 3D object surface mesh can be computed since both color and depth is available. This object mesh is later used for interaction with the computer generated elements. By merging the object with the background scene model, we have a full 3D reconstruction of the real scene geometry at hand, even the occluded background behind the object.

2.5 Moving the Camera Head

Free movement of the camera is a prerequisite for a versatile virtual studio. Virtual studios with provision to move the camera exist but mostly rely on complex and expensive camera tracking devices [6]. In our approach, we can relax these requirements by exploiting the known environment model. Köser et.al. [23] propose to use visual tracking of a previously constructed environment model with the help of a fish-eye camera. Hereby the tracking does not rely on any artificial markers, since the model itself is used as 3D reference system. The wide field of view of the fish-eye lens will show large parts of the tracked model, even if the camera moves quickly or a dynamic object is occluding parts of the scene. We follow this approach and apply fish-eye camera pose tracking within the environment. The camera tracking can be sustained over long sequences without drift, because the 3D environment model does not change over time.

3 Real-Time 3D Interaction

The previous section described the components of the system. In this section we will discuss the interaction capabilities.

The ability to combine virtual content with real image footage using keying and camera parameter tracking already extends the possibilities of virtual studios. However, without a 3D reconstruction of the real part, an interaction between the real and virtual content is difficult to establish. Regarding [24], the most important optical interactions, which significantly improve the augmentation, are occlusions, shadow casting and reflections. It is possible to use a chroma keyer to segment a real object's shadows or to key the reflection of the real object from a shiny surface, but this requires to physically model the real surface. This might be feasible for floors and walls, however for more complicated objects it is normally too expensive to manually construct a virtual model and a physical counterpart for capturing the shadows and reflections, even more so if dynamic virtual content shall be used.

Furthermore, it is not possible to automatically determine mutual occlusions between the real and virtual content without extracting their relative depth distribution w.r.t. the augmented view. Since the content is dynamic and should be reconstructed in real-time, approaches using passive stereoscopy from images and laser scanners are at their limits. In virtual studios, shape-from-silhouette algorithms are used due to their stability and speed, with multiple cameras capturing different views of the dynamic real content. Chroma keying is used to retrieve view-dependent silhouettes, which are combined to a 3D visual hull. Although this hull is not as detailed as laser scans, it often suffices to handle occlusion and to integrate shadow casting. [24] gives an overview on the limitations and variants of this approach. For good results, multiple wide baseline viewpoints[5] have to be calibrated, captured and evaluated simultaneously. An alternative is provided by the use of ToF-Cameras. As discussed in [25,26] the

---

[5] Following [24] 6 up to 12 cameras are required to deliver reasonable results.

depth can be used to handle mutual occlusions without complex algorithmic effort, even without the use of multiple viewpoints. Although simple in its construction and handling, a system using a single viewpoint is only able to deliver a 2.5D model of the reconstructed object. This restricts the scenarios in which correct shadow and reflection calculation can be performed. An approach for visual hull calculation using multiple ToF and color cameras is presented in [27].

We will exploit our proposed system to handle occlusions, shadows, and reflections. Most of these tasks can be performed in real-time or near real-time, allowing instantaneous feedback between director and actor during shooting. More advanced interactions could be added in a post production phase as the full 3D geometry is at hand.

3.1   Mutual Occlusions between Background, Objects and CG Elements

The computation of mutual occlusions between computer generated elements and background, and between the dynamically moving person and computer generated elements is straight forward with our approach. The depth keying delivers all necessary information to compute pixel-accurate depth for the moving object and the background, and a virtual object will be either occluded by the person or will occlude the person and the background, depending on the relative depth w.r.t. the viewing camera. Since the 3D environment was scanned with true 3D metrics, one can locate the computer generated objects at the correct position and with correct metric size without any problem. The objects are placed onto the floor by simply dropping them, and gravity and collision detection with the modeled floor will automatically put the object in place. The camera view is then augmented by color mixing of the computer generated elements into the image at positions where the computer generated objects are not occluded. The rest of the image is taken from the current color image, so consistency is guaranteed, like the correct shadows of the real person on the real walls. Figure 5 shows an example of such depth-based color mixing. Note the mutual occlusions: the statue is occluding the background while being occluded by the real person, which again is occluded by the artificial plants. In the occluding case, a very precise depth segmentation is crucial, and measurement noise might degrade the segmentation. This is also due to the fact that currently our depth camera delivers 12.5 fps while the color camera is running with 30 fps. This is a technical problem which will be solved in the near future, since better and synchronized depth cameras are already announced[6]. The calibration/registration error is well below one pixel so a precise pixel mapping between depth and color camera ist achieved using the mentioned approaches.

Occlusions may appear, but as the baseline of the cameras is small compared to the scene distance occlusions are small. Alternatively 2D/3D-cameras are

---

[6] New ToF-cameras are announced which allow full synchronized depth images at 25 fps.

Fig. 5. Color and depth mixing for a frame of an input sequence. Left: Original image of person walking, augmented by virtual objects with mutual occlusion. Right: corresponding depth map used for mixing.

already in development which use a single lens and a beam splitter for high resolution intensity and depth information. Using this new camera occlusions will no longer be an issue.

3.2   Light and Shadow Casting

While correct depth keying is the key to proper occlusion handling, proper lighting is important for realistic appearance of the computer generated objects. Therefore, also the lights must be modeled accordingly so that the computer generated elements are lit similar to the real scene objects. Furthermore, the computer generated elements must cast realistic shadows onto the elements of the real scene (background and person) and vice versa. Correct shadows are of particular importance for virtual objects placed inside a real scene, since without correct shadow on the floor, the object seems to float in space.

Light interaction is possible in our system once we add a model of the real light sources and light source position. Currently, the light source model is defined manually and the positions of the lights are selected in the background model by hand, but automated detection of the light source's position is not difficult. For example, the linear light arrays in the ceiling are already part of the model geometry and can be detected easily. Once the light geometry, fraction and

Fig. 6. Shadow casting. Left: mixed image with added shadows, right: shadow map for two light sources.

temperature is defined, the geometric scene model allows to cast mutual shadows. Note that the light sources are only approximated by point lights for simplicity and real-time capability.

To add the shadows, which are cast by virtual objects onto the real images, light maps are calculated for each video frame. These maps basically encode how much light is reaching a particular pixel of the image when virtual content is present. Each pixel in the light map contains a factor $0 \leq s \leq 1$, which is used to scale the RGB color values in the respective augmented image. A scale factor of 1 corresponds to no shadowing, 0 renders a pixel absolutely black and values in between model partial soft shadowing. The light map operates on the 2D image and reduces only those image parts, which are visible to the user but shaded by an object.

The light maps are generated using the shadow mapping technique [28]. For each light source, a depth map is rendered for all objects that cast shadows. These are the computer generated objects as well as the dynamically moving foreground person. Next, the background model and all (real and virtual) objects are rendered from the camera's point of view, shading the scene with the calculated lights' depth maps using projective texturing.

This way, for each pixel in the image the distance values R encoded in the light sources' depth maps can be compared to the distances D between a light source and the 3D point corresponding to the pixel. As the light's depth map provides us with the distance between the light source and the first intersection of the light ray with the scene geometry, we can decide whether the pixel is in shadow ($R < D$) or receives light from the light source ($D = R$). Evaluating all light sources and combining them with an ambient light offset yields the view dependent light map used for shadow generation as shown in figure 6. The light maps are additionally filtered with a Gaussian filter to soften the shadows but no real soft shadows are used due to real-time demands.

One exception is made for the interaction between the real foreground and background object. In the target image, the real light source already casts a shadow of the real person onto the background. Hence, the shadow test between real object and background is disabled, while the shadow casting between the real and virtual objects is computed. Figure 6 demonstrates this mutual shadowing. The virtual shadow of the statue of liberty, which is cast onto the back wall, is consistent with the real shadow of the person. Also, the person casts a shadow onto the statue and vice versa. In figure 6 (right), the computed shadow masks can be seen. Only the image areas that are dimmed by the shadows are marked, hence the foreground image region of the statue and the person is left untouched. The problem of double shadows remains, as visible in figure 6 at the bottom left, where the shadows of the Statue of Liberty and of the person are superimposed and doubling each other at the background. This is a known issue and not addressed by our approach. In literature methods are described (e.g. in [29] and [30]) which can be used to solve that issue.

Fig. 7. Object reflection. Right: Dinosaur without (top) and with (bottom) correct surface reflection in the environment model. Left: Reflective marble floor that reflects environment and real person model.

3.3   Surface Reflections

Another cue that is important for correct visual appearance is the reflection of a (real or virtual) object on a shiny surface (see figure 7). A subtle example of this reflection is found on the table surfaces of the background model. A little clay dinosaur is placed onto the table to the right by dropping it there. When looking at the image, something is wrong, since the dinosaur seems to float in space (see figure 7, top right), but there are no visible shadow cues as the light comes from the front. Closer inspection reveals that the table top is slightly reflective, as can be seen by the reflection of the AC current plug and cable on the table. Hence, a reflection of the dinosaur will remedy the problem. In our system, such reflection is easily incorporated, since we know the correct surface normal of the surface, the camera and light viewing direction, and we have a complete 3D environment which is mirrored in the surface. Figure 7 (bottom right) shows the effect of adding the reflection for the dinosaur, where the reflectivity was tuned by hand for this example. Of course, this property can also be exploited to purposely insert mirrored surfaces. In figure 7 (left) we inserted an artificial floor patch with some highly reflective marble surface. Both, environment model and the dynamic person model, are reflected correctly. This reflection pushes the approach to the limit, since it renders the dynamic object from a very different perspective. Even slight depth errors will produce gross reflection errors, so highest depth-quality is needed.

4   Discussion and Future Directions

Based on a Time-of-Flight depth camera, coupled with color cameras onto a pan-tilt head, we presented a mobile and flexible system for mixed reality applications: MixIn3D. After systematically scanning the environment to set up a background model, the system can be used for keying and occlusion determination between real and virtual objects, for shadowing and reflections.

The major challenge of the current system is to handle correct depth segmentation from the low-resolution depth data of the Time-of-Flight camera. There is much ongoing research and development activity, and improved cameras are already announced. Thus, we are convinced that this problem will be solved in the near future. Also, combining the ToF-data with additional vision-based segmentation algorithms will likely improve the quality further.

The main advantage of MixIn3D, as compared to other mixed reality systems, is that all data is fully available in 3D, so that more complicated interactions are possible - if not in real-time then at least in a post-processing step.

References

1. Xu, Z., Schwarte, R., Heinol, H., Buxbaum, B., Ringbeck., T.: Smart pixel - photonic mixer device (PMD). In: M2VIP 1998 - International Conference on Mechatronics and Machine Vision in Practice, pp. 259–264 (1998)

2. Lange, R., Seitz, P., Biber, A., Schwarte, R.: Time-of-Flight range imaging with a custom solid-state imagesensor. In: EOS/SPIE Laser Metrology and Inspection, vol. 3823 (1999)
3. Thomas, G.: Mixed reality techniques for TV and their application for on-set and pre-visualization in film production. In: International Workshop on Mixed Reality Technology for Flimmaking (2006)
4. Smith, A.R., Blinn, J.F.: Blue screen matting. In: SIGGRAPH 1996: Proceedings of the 23rd annual conference on Computer graphics and interactive techniques, pp. 259–268. ACM, New York (1996)
5. Wang, J., Cohen, M.F.: Image and video matting: a survey. Found. Trends. Comput. Graph. Vis. 3(2), 97–175 (2007)
6. Thomas, G.A., Jin, J., Niblett, T., Urquhart, C.: A versatile camera position measurement system for virtual reality. In: Proceedings of International Broadcasting Convention, pp. 284–289 (1997)
7. Streckel, B., Koch, R.: Lens model selection for visual tracking. In: Kropatsch, W.G., Sablatnig, R., Hanbury, A. (eds.) DAGM 2005. LNCS, vol. 3663, pp. 41–48. Springer, Heidelberg (2005)
8. Chandaria, J., Thomas, G., Bartczak, B., Koeser, K., Koch, R., Becker, M., Bleser, G., Stricker, D., Wohlleber, C., Felsberg, M., Hol, J., Schoen, T., Skoglund, J., Slycke, P., Smeitz, S.: Real-time Camera Tracking in the MATRIS Project. In: Proceedings of International Broadcasting Convention (IBC), Amsterdam, The Netherlands, pp. 321–328 (2006)
9. Schiller, I., Beder, C., Koch, R.: Calibration of a PMD camera using a planar calibration object together with a multi-camera setup. In: Proceedings of the ISPRS Congress, Bejing, China (2008)
10. Lindner, M., Schiller, I., Kolb, A., Koch, R.: Time-of-Flight Sensor Calibration for Accurate Range Sensing. In: Computer Vision and Image Understanding CVIU, Special Issue on Time-of-Flight Camera based Computer Vision (2009) (accepted for publication)
11. Zhang, Z.: Flexible Camera Calibration by Viewing a Plane from Unknown Orientations. In: Proceedings of the International Conference on Computer Vision, Corfu, Greece, pp. 666–673 (1999)
12. Bouguet, J.: Visual methods for three-dimensional modelling. PhD thesis, California Institute of Technology (1999)
13. Scaramuzza, D., Martinelli, A., Siegwart, R.: A Flexible Technique for Accurate Omnidirectional Camera Calibration and Structure from Motion. In: Proceedings of IEEE International Conference of Vision Systems. IEEE, Los Alamitos (2006)
14. Fuchs, S., May, S.: Calibration and Registration for Precise Surface Reconstruction with TOF Cameras. In: Proceedings of the DAGM Dyn3D Workshop, Heidelberg, Germany (2007)
15. Lindner, M., Kolb, A.: Lateral and Depth Calibration of PMD-Distance Sensors. In: Bebis, G., Boyle, R., Parvin, B., Koracin, D., Remagnino, P., Nefian, A., Meenakshisundaram, G., Pascucci, V., Zara, J., Molineros, J., Theisel, H., Malzbender, T. (eds.) ISVC 2006. LNCS, vol. 4292, pp. 524–533. Springer, Heidelberg (2006)
16. Huhle, B., Jenke, P., Straer, W.: On-the-fly scene acquisition with a handy multisensor-system. In: Dynamic 3D Imaging Workshop in Conjunction with DAGM 2007 (2007)
17. Scharstein, D., Szeliski, R., Zabih, R.: A Taxonomy and Evaluation of Dense Two-frame Stereo Correspondence Algorithms. In: Proceedings of IEEE Workshop on Stereo and Multi-BaselineVision, Kauai, HI (December 2001)

18. Seitz, S., Curless, B., Diebel, J., Scharstein, D., Szeliski, R.: A Comparison and Evaluation of Multi-View Stereo Reconstruction Algorithms. In: Proceedings of International Conference Computer Vision and Pattern Recognition, CVPR (2006)
19. Grundhöfer, A., Bimber, O.: VirtualStudio2Go: digital video composition for real environments. ACM Trans. Graph. 27(5), 1–8 (2008)
20. Garland, M., Heckbert, P.S.: Surface simplification using quadric error metrics. In: SIGGRAPH 1997, pp. 209–216 (1997)
21. Anatol, F., Falko, K., Bogumil, B., Reinhard, K.: Generation of 3D-TV LDV-Content with Time-of-Flight Camera. In: 3DTV Conference: The True Vision - Capture, Transmission and Display of 3D Video, Potsdam, Germany, May 2009, pp. 1–4 (2009)
22. Yang, Q., Yang, R., Davis, J., Nister, D.: Spatial-depth Super Resolution for Range Images. In: Proc. IEEE Conference on Computer Vision and Pattern Recognition CVPR 2007, pp. 1–8 (2007)
23. Koeser, K., Bartczak, B., Koch, R.: Robust GPU-Assisted Camera Tracking using Free-form Surface Models. Journal of Real Time Image Processing 2, 133–147 (2007)
24. Grau, O.: 3D in Content Creation and Post-Production. In: Schreer, O., Peter Kauff, T.S. (eds.) 3D Videocommunication, pp. 39–53 (2005)
25. Gvili, R., Kaplan, A., Ofek, E., Yahav, G.: Depth keying. SPIE, vol. 5006, pp. 564–574 (2003)
26. Bartczak, B., Schiller, I., Beder, C., Koch, R.: Integration of a Time-of-Flight Camera into a Mixed Reality System for Handling Dynamic Scenes, Moving Viewpoints and Occlusions in Real-Time. In: Proceedings of the 3DPVT Workshop, Atlanta, GA, USA (June 2008)
27. Guan, L., Franco, J.S., Pollefeys, M.: 3D Object Reconstruction with Heterogeneous Sensor Data. In: Proceedings of the 3DPVT Workshop, Atlanta, GA, USA (June 2008)
28. Williams, L.: Casting curved shadows on curved surfaces. In: SIGGRAPH 1978: Proceedings of the 5th annual conference on Computer graphics and interactive techniques, pp. 270–274. ACM, New York (1978)
29. Jacobs, K., Nahmias, J.D., Angus, C., Reche, A., Loscos, C., Steed, A.: Automatic generation of consistent shadows for augmented reality. In: GI 2005: Proceedings of Graphics Interface 2005, School of Computer Science, University of Waterloo, Waterloo, Ontario, Canada, Canadian Human-Computer Communications Society, pp. 113–120 (2005)
30. Gibson, S., Cook, J., Howard, T., Hubbold, R.: Rapid shadow generation in real-world lighting environments. In: EGRW 2003: Proceedings of the 14th Eurographics workshop on Rendering, Aire-la-Ville, Switzerland, Eurographics Association, pp. 219–229 (2003)

Self-Organizing Maps for Pose Estimation with a Time-of-Flight Camera

Martin Haker, Martin Böhme, Thomas Martinetz, and Erhardt Barth

Institute for Neuro- and Bioinformatics, University of Lübeck
Ratzeburger Allee 160, 23538 Lübeck, Germany
{haker,boehme,martinetz,barth}@inb.uni-luebeck.de
http://www.inb.uni-luebeck.de

Abstract. We describe a technique for estimating human pose from an image sequence captured by a time-of-flight camera. The pose estimation is derived from a simple model of the human body that we fit to the data in 3D space. The model is represented by a graph consisting of 44 vertices for the upper torso, head, and arms. The anatomy of these body parts is encoded by the edges, i.e. an arm is represented by a chain of pairwise connected vertices whereas the torso consists of a 2-dimensional grid. The model can easily be extended to the representation of legs by adding further chains of pairwise connected vertices to the lower torso. The model is fit to the data in 3D space by employing an iterative update rule common to self-organizing maps. Despite the simplicity of the model, it captures the human pose robustly and can thus be used for tracking the major body parts, such as arms, hands, and head. The accuracy of the tracking is around 5–6 cm root mean square (RMS) for the head and shoulders and around 2 cm RMS for the head. The implementation of the procedure is straightforward and real-time capable.

1 Introduction

A time-of-flight (TOF) camera [1] provides a range map that is perfectly registered with an intensity image (often referred to as an *amplitude* image in TOF nomenclature), making it an attractive sensor for a wide range of applications.

In this paper, we present a technique for estimating human pose in 3D based on a simple model of the human body. The model consists of a number of vertices that are connected by edges such that the resulting graph structure resembles the anatomy of the human body, i.e. the model represents the torso, the head, and the arms. The model is updated using an iterative learning rule common to self-organizing maps (SOMs) [2]. The position of certain body parts, such as the hands, can be obtained from the model as the 3D coordinates of the corresponding vertices, i.e. the position of the hands in 3D corresponds to the position of the vertex that terminates the chain representing an arm. Thus, body parts can be tracked in 3D space.

The estimation of 3D human pose has been addressed in a number of different publications. The majority of work focuses on the estimation of pose from

R. Koch and A. Kolb (Eds.): Dyn3D 2009, LNCS 5742, pp. 142–153, 2009.
© Springer-Verlag Berlin Heidelberg 2009

single images taken with a regular 2D camera, and a number of different algorithmic approaches have been presented. In [3] the pose is recovered from shape descriptors of image silhouettes. The authors of [4] map low-level visual features of the segmented body shape to a number of body configurations and identify the pose as the one corresponding to the most likely body configuration given the visual features. An approach based on a large database of example images is presented in [5]. The authors learn a set of parameter-sensitive hashing functions to retrieve the best match from the database in an efficient way.

Very accurate 3D reconstruction of human motion from multi-view video sequences was published in [6]. Based on a segmentation of the subject, the authors use a multi-layer framework that combines stochastic optimization, filtering, and local optimization to estimate the pose using a detailed model of the human body. However, the computational cost is relatively high and the system does not operate at camera frame rates.

Pose estimation based on 3D data has been addressed in [7]. The 3D volume of a person is estimated in a multi-camera setup using the shape-from-silhouette method. A skeleton model is then fit to a 2D projection of the volumetric data. The 2D projection is obtained by a virtual camera and the model is fit using certain features of the outer contour. The 3D coordinates of the model are finally reconstructed by inverting the 2D projection of the virtual camera, i.e. the vertices of the skeleton are projected back into 3D space using the intrinsic parameters of the virtual camera.

Another approach to obtaining a skeleton in 3D is to apply a thinning to volumetric data directly in 3D space [8,9]. The human pose can then be estimated from the skeleton [10].

Two related methods based on stereo imaging were presented in [11] and [12]. The authors introduce a hierarchical human body model database. For a given image the algorithm uses both silhouette and depth information to identify the model pose with the best match.

The work in [13] fuses 2D and 3D information obtained from a stereo rig and a TOF camera to fit a human body model composed of generalized cylinders. The system models body joints and uses kinematic constraints to reduce the degrees of freedom. The 3D data is obtained using a TOF camera and the system runs at frame rates of 10–14 frames per second.

Another recent approach using TOF cameras was presented in [14]. The method tracks a number of anatomical landmarks in 3D over time and uses these to estimate the pose of an articulated human model. The model is in turn used to resolve disambiguities of the landmark detector and to provide estimates for undetected landmarks. The entire approach is very detailed and models constraints such as joint limit avoidance and self-penetration avoidance. Despite its complexity, the method runs at a frame rate of approximately 10 frames per second.

Our approach, in contrast, is a very simple one that demonstrates how effectively TOF cameras can be used to solve relatively complex computer vision tasks. A general advantage of TOF cameras is that they can provide both range and intensity images at high frame rates. The combined use of both types of data was already used for tracking [15,16] and allows a robust segmentation of the human body in front of the camera. The range data, representing a $2\frac{1}{2}$D image, can then be used to obtain a point cloud in 3D representing the visible surface of the person. Thus, limbs extended towards the camera can still be easily identified while this proves to be a more difficult task in 2D projections of a scene.

Our approach takes advantage of this property and fits a simple model of the human body into the resulting point cloud in 3D. The model fitting algorithm is based on a SOM, can be implemented in a few lines of code, and the method runs at frame rates up to 25 frames per second on a 2 GHz Intel Core 2 Duo. The algorithmic approach to this procedure is discussed in Sect. 2. The method delivers a robust estimation of the human pose, as we show in Sect. 3 for image data that was acquired using a MESA SR4000 TOF camera.

2   Method

The first step of the proposed procedure is to segment the human body from the background of the image. We employ a simple thresholding approach that uses both range and intensity data. The thresholds for the two images are determined adaptively for each frame.

In case of the amplitude image, the pixel values correspond to the amount of light of the TOF camera's active illumination that is reflected back into the camera. Hence, the amplitude can be considered a confidence measure for the accuracy of the range measurement because it indicates the measurement's signal-to-noise ratio. The attenuation of the amplitude is proportional to the squared distance of an object to the camera. Thus, objects close to the camera appear generally much brighter than the background. We use the Otsu threshold [17] to determine an adaptive value for the threshold that separates the dark background from the brighter foreground. A more accurate segmentation using thresholding on amplitude data proves to be difficult because the objects may have different properties of reflecting infrared light.

In case of the range data, this simple assumption of a bimodal distribution does not hold if multiple objects are located at different distances in front of the camera. Thus, we construct a histogram of the range values in which every object can be assumed to result in a peak if the objects are truly at different distances from the camera. The threshold is determined as the one that separates the peak corresponding to the closest object from the peaks of the remaining objects.

The final segmented image is obtained as the one where the foreground pixels have been classified as foreground pixels with respect to both types of data. Furthermore, we identify the largest connected component of foreground pixels and consider all remaining pixels background. Thus, we obtain a clear segmentation of a single person closest to the camera in most cases. A sample TOF image and the resulting segmented image is shown in Fig. 1.

The identified foreground pixels can be assumed to sample the visible surface of the person in front of the camera. Since the intrinsic parameters of the

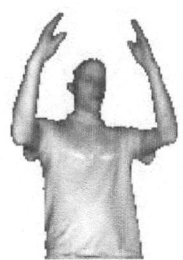

Fig. 1. Sample image taken with a MESA SR4000 TOF camera. The leftmost image shows the amplitude data. The range image is given in the center and the resulting segmentation is shown on the right.

camera, such as focal length and pixel size, are known, the surface pixels can be projected back into 3D space. As a result one obtains a point cloud in 3D that represents the 3-dimensional appearance of the person. This approach has two major advantages: (i) The representation is scale-invariant due to the fact that the size of the person in 3D space remains the same independently of the size of its image; (ii) body parts that are extended towards the camera in front of the torso can be easily identified due to the variation in distance, whereas this information is lost in 2D projections of the scene obtained with regular cameras.

Our method aims at fitting a simple graph model representing the anatomy of the human body into the resulting point cloud in 3D. To this end, we employ a SOM. We define a graph structure of vertices and edges that resembles a frontal view of the human body. Body parts, such as arms and torso, are modeled by explicitly defining the neighborhood structure of the graph, i.e. an arm is represented by a simple chain of pairwise connected vertices whereas vertices in the torso are connected to up to four neighbors forming a 2D grid. The resulting model structure is depicted in Fig. 2.

The SOM is updated by an iterative learning rule for each consecutive frame of the video sequence. The first frame uses the body posture depicted in Fig. 2 as an initialization of the model. During initialization the model is translated to the center of gravity of the 3D point cloud. The scale of the model is currently set manually to a fixed value that corresponds to an average-sized person. We can report that the scale is not a particularly critical parameter and that the same fixed scale works for adults of different height. Once the scale is set to an appropriate value, there is no need to adjust it during run-time due to the above mentioned scale-invariance of the method. The update of the model for each consecutive frame then depends on the model that was estimated for the previous frame.

The adaptation of the model to a new frame involves a complete training of the SOM, i.e. a pattern-by-pattern learning is performed using the data points of the 3D point cloud. This iterative procedure selects a sample vector x from the point cloud at random and updates the model according to the following learning rule:

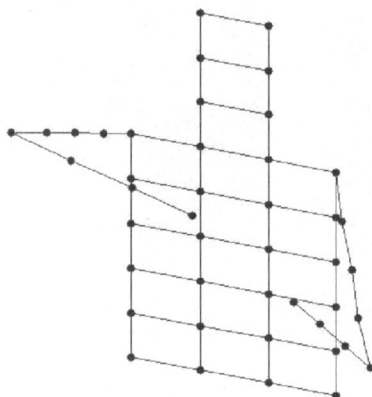

Fig. 2. Graph model of the human body. The edges define the neighborhood structure for the SOM.

$$\hat{v}^{t+1} = \hat{v}^t + \hat{\epsilon}^t \cdot (x - \hat{v}^t) \tag{1}$$
$$\tilde{v}^{t+1} = \tilde{v}^t + \tilde{\epsilon}^t \cdot (x - \tilde{v}^t). \tag{2}$$

Here, $\hat{v}$ denotes the node that is closest to the sample x with respect to the distance measure $d(x, v) = \|x - v\|_2$. The nodes $\tilde{v}$ are the neighbors of $\hat{v}$ as defined by the model structure. The learning rates are denoted by $\hat{\epsilon}^t$ and $\tilde{\epsilon}^t$ for the closest node and its neighbors, respectively. The learning rate $\hat{\epsilon}^t$ was set to:

$$\hat{\epsilon}^t = \epsilon_i \cdot (\epsilon_f/\epsilon_i)^{t/t_{\max}}. \tag{3}$$

Here, $t \in \{0, \ldots, t_{\max}\}$ denotes the current adaptation step for this frame and $t_{\max}$ denotes the total number of adaptation steps performed for this frame. The initial learning rate ϵ_i and the final learning rate ϵ_f were set to 0.1 and 0.05. The learning rate for the neighbors was chosen to be $\tilde{\epsilon}^t = \hat{\epsilon}^t/2$. This choice of the learning rate was already proposed in previous work on self-organizing networks [18]. The initial and final learning rates were set to relatively high values in order to allow the network to handle fast movements of the person, i.e. if the limbs are moved quickly the correctional updates for the corresponding nodes have to be large so that the model can accurately follow.

This update rule does not always guarantee that the topology of the model is preserved. Here, we refer to topology with respect to the connectivity of the nodes within body parts such as the arm. Imagine the situation where the subject's hands touch in front of the torso. If the hands are separated again, it is possible that the model uses the last node of the left arm to represent samples that actually belong to the hand of the right arm. It can thus happen, that the last node of the left arm may continue to be attracted by the right hand although both hands have moved apart and, thus, the left arm will extend into empty space. In principle, the update rules resolve this problem over time. However, only a small number of updates are performed per frame and this may lead to a wrong estimation of the topology for a small number of frames.

To avoid this, we developed a modification of the update rule that speeds up the learning process by forcing neighboring vertices to stay close together. This is achieved by the following rule that is applied after the actual learning step if the distance $d(\hat{\boldsymbol{v}}, \tilde{\boldsymbol{v}}_a)$ exceeds a certain threshold θ:

$$\hat{\boldsymbol{v}} = \tilde{\boldsymbol{v}}_a + \theta \cdot \frac{(\hat{\boldsymbol{v}} - \tilde{\boldsymbol{v}}_a)}{\|\hat{\boldsymbol{v}} - \tilde{\boldsymbol{v}}_a\|_2}. \tag{4}$$

Here, $\tilde{\boldsymbol{v}}_a$ is a specific neighbor of $\hat{\boldsymbol{v}}$ referred to as an anchor. The rule enforces that the distance between the vertex $\hat{\boldsymbol{v}}$ and its anchor is always less than or equal to θ. The threshold θ depends on the scale of the model. The anchor of each vertex is defined as the neighbor that has minimal distance to the center of the torso with respect to the graph structure of the model, i.e. it is the vertex that is connected to the center of the torso by the smallest number of edges.

3   Results

3.1   Qualitative Evaluation

The proposed method was evaluated using a MESA SR4000 TOF camera. We operate the camera at a modulation frequency of 30 MHz for the active illumination. As a result the camera can disambiguate distances in the range of up to 5 meters. In the following sample images, the person was at a distance of roughly 2.5 meters from the camera. At that distance the range measurement has an accuracy of approximately 1 cm.

A sample result of the pose estimation is shown in Fig. 3. The figure depicts the point cloud of samples in 3D that represent the visual surface of the person in front of the camera shown in Fig. 1. The model that was fitted to the point cloud is imprinted into the data. One can observe that the model captures the anatomy of the person correctly, i.e. the torso is well covered by the 2-dimensional grid, a number of vertices extend into the head, and the 1-dimensional chains of vertices follow the arms. Thus, the position of the major body parts, such as the hands, can be taken directly from the corresponding vertices of the model in 3D.

The data from Fig. 3 is taken from a sequence of images. Further sample images from this sequence are given in Fig. 4. Each image shows the segmented amplitude image with the imprinted 2D projection of model. One can observe that the model follows the movement of the arms accurately, even in difficult situations where the arms cross closely in front of the torso. Note that the procedure does not lose the position of the head even though it is occluded to a large extent in some of the frames. The sample images are taken from a video sequence, which is available under http://www.artts.eu/demonstrations/

It is important to point out that the method may misinterpret the pose. This can for example be the case if the arms come too close to the torso. In such a case the SOM cannot distinguish between points of the arm and the torso within the 3D point cloud. We can report, however, that the method can recover the true configuration within a few frames once the arms are extended again in most cases.

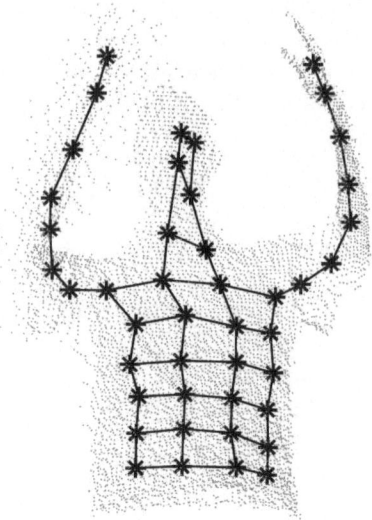

Fig. 3. Point cloud sampling the visible surface of a human upper torso in 3D. The graph represents the human model that was fitted to the data.

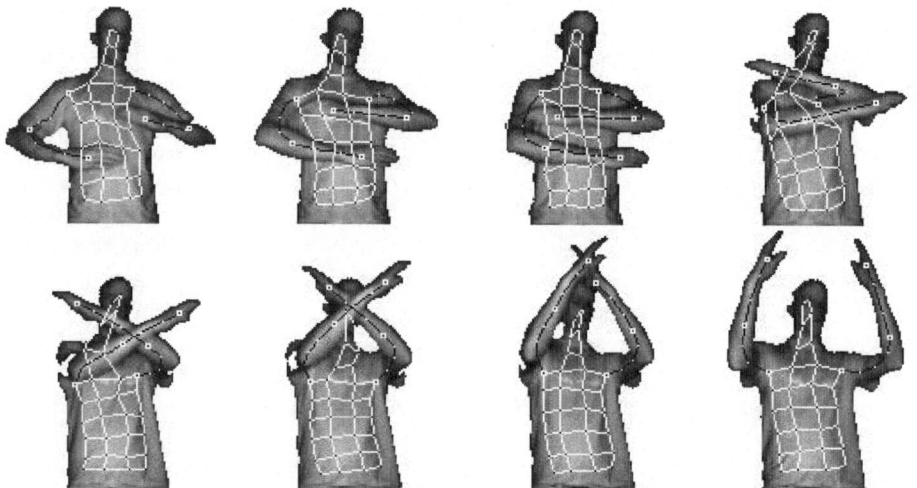

Fig. 4. A selection of frames from a video sequence showing a gesture. The model estimated by the pose estimation is imprinted in each frame. The edges belonging to torso and head are colored in white, whereas the arms are colored in black.

We assume that it is possible to detect and avoid such problems by imposing a number of constraints on the model, e.g. that the arms may only bend at the elbows and that the entire model should generally be oriented such that the head is pointing upwards. However, note that the current results were achieved without any such constraints.

3.2   Quantitative Evaluation

To evaluate the accuracy of the tracking quantitatively, we acquired sequences of 5 persons moving in front of the camera; each sequence was around 140 to 200 frames long. In each frame, we hand-labeled the positions of five parts of the body: the head, the shoulders, and the hands. To obtain three-dimensional ground-truth data, we looked up the distance of each labeled point in the range map and used this to compute the position of the point in space. This implies that the sequences could not contain poses where any of the body parts were occluded; however, many poses that are challenging to track, such as crossing the arms in front of the body, were still possible, and we included such poses in the sequences. (Note that the tracker itself can track poses where, for example, the head is occluded; see Fig. 4.)

The labeled positions can now be compared with the positions of the corresponding nodes in the tracked model. However, when assessing the accuracy of the tracking in this way, we run into the problem that we never define explicitly which part of the body each node should track. For example, though the last node in each of the arms will typically be located on or near the hand, we do not know in advance exactly which part of the hand the node will track. This means that there may be a systematic offset between the position that is labeled as "hand" and the position that the hand node tracks. To give a realistic impression of tracking accuracy, we should eliminate these systematic offsets.

We do this by measuring the average offset between the tracked position and the labeled position on ten "training" frames; this offset is then used to correct the tracked position in the remaining "test" frames, on which the accuracy is measured. Because the orientation of the respective parts of the body can change, we need to measure the offsets not in the world coordinate system but in a local coordinate system. For the head and shoulders, we use a coordinate system where the x-axis points from the left shoulder (of the tracked model) to the right shoulder, the y-axis is defined so that the head lies in the x-y-plane, and the z-axis is perpendicular to the other two axes to form a right-handed coordinate system. For the hands, it is not as easy to define a full coordinate system because the model only measures the direction in which the forearm is pointing but not the orientation of the hand. For this reason, we estimate and correct the offset between tracked and labeled position only along the direction of the forearm, which we define by the last two nodes in the arm; this is the direction that accounts for most of the offset. Any offset perpendicular to the direction of the forearm is not corrected.

Once the tracked positions have been corrected in this way, we can measure the tracking error. Fig. 5 shows a plot of tracking error over time for one of

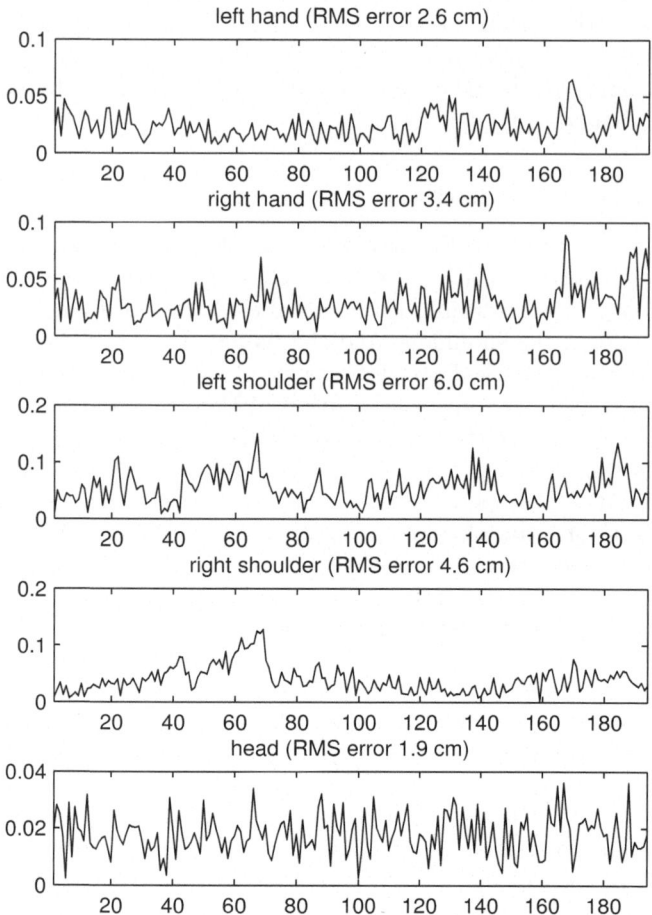

Fig. 5. Plots of tracking error over time for one of the sequences. The horizontal axis plots frame number, the vertical axis plots tracking error in meters.

the recorded sequences. It is obvious that there is little to no systematic error remaining; instead, most of the error is due to tracking noise.

Table 1 shows the root mean square (RMS) tracking error, averaged over all frames and subjects. The average error is around 5 to 6 cm for the hands and shoulders and around 2 cm for the head. While this degree of accuracy is not sufficient for tracking very fine movements, it is more than adequate for determining overall body posture and for recognizing macroscopic gestures. Also, consider that no smoothing of the tracked positions over time was carried out.

A major advantage of the proposed method is that the training of the model converges very fast for each new frame. Thus, only a small number of the samples of the 3D cloud need actually be considered during the update even when the person performs very fast movements in front of the camera. The sample image

Table 1. Root mean square (RMS) error between tracked and labeled positions, averaged over all frames and subjects

body part	RMS error
left hand	5.90 cm
right hand	5.29 cm
left shoulder	5.32 cm
right shoulder	5.15 cm
head	2.21 cm

in Fig. 1 contains roughly 6500 foreground pixels. However, we use only 10% of these samples for updating the model, i.e. we select roughly 650 points in 3D in random order from the point cloud and use these for updating the model by pattern-by-pattern learning. As a result the computational complexity is very low, and we achieve frame rates up to 25 frames per second on a 2 GHz PC while robustly tracking the human pose in scenarios such as the one depicted in Fig. 4. The use of a higher number of samples for training will further increase the robustness while at the same time the frame rate will decrease.

4   Discussion

We have presented a simple procedure to estimate human pose from a sequence of range images. The procedure is especially suitable for TOF cameras as they can deliver range data in combination with intensity images at high frame rates. These cameras can be assumed to be available at relatively low costs in the near future.

The use of a SOM results in a very simple, yet very efficient implementation. In principle the procedure can be extended easily to any other kind of deformable object.

A major shortcoming of the current implementation is that the method cannot deal with multiple persons in front of the camera, i.e. the system always assumes that the segmented foreground pixels correspond to a single person. This approach fails for example if two people are at the same distance in front of the camera and very close together. In that case the segmented foreground pixels sample the visual surface of both persons. Since the SOM attempts to represent all samples equally the resulting pose estimation fails. Using the current approach, this problem must be solved by an improved method for segmentation that can handle multiple objects in front of the camera. Then, a SOM can be trained for each segmented object and thus multiple people can be tracked.

This in turn can lead to a related problem that occurs when the segmentation fails to detect parts of the body due to occlusion, e.g. when the lower part of an arm is occluded by a second person. In that case the SOM will use the entire chain of arm nodes to represent the upper part of the arm. Thus, the node for the hand will be misplaced. To tackle this problem the system needs to identify the presence of certain body parts based on pose estimates from previous frames. In case occluded body parts have been identified, the corresponding nodes of the SOM must be excluded from training. Instead their location could be predicted based on the posture of the remaining model. These two issues need to be addressed in future work.

There exist other approaches that compute a more accurate estimate of human pose but our goal within this work was to develop a simple method that gives a rough but robust estimate of human pose at high frame rates.

We intend to use the proposed method of pose estimation for action recognition and gesture-based man-maschine interaction. Generally, the evaluation of certain spatio-temporal features for the analysis of video sequences is computationally expensive. We argue that rough knowledge of the position of landmarks, such as the hands, can greatly improve the runtime of feature-based action recognition systems, because the features do not have to be evaluated over the entire video sequence but only at those locations where certain important landmarks have been detected. Furthermore, these features can be put into a larger context if their relative location to each other is known.

Acknowledgments

This work was developed within the ARTTS project (www.artts.eu), which is funded by the European Commission (contract no. IST-34107) within the Information Society Technologies (IST) priority of the 6th Framework Programme. This publication reflects the views only of the authors, and the Commission cannot be held responsible for any use which may be made of the information contained therein.

References

1. Oggier, T., Büttgen, B., Lustenberger, F., Becker, G., Rüegg, B., Hodac, A.: SwissRanger™ SR3000 and first experiences based on miniaturized 3D-TOF cameras. In: Ingensand, K. (ed.) Proc. 1st Range Imaging Research Day, Zurich, pp. 97–108 (2005)
2. Kohonen, T.: Self-Organizing Maps. Springer, Berlin (1995)
3. Agarwal, A., Triggs, B.: Recovering 3D human pose from monocular images. IEEE Transactions on Pattern Analysis and Machine Intelligence 28(1), 44–58 (2006)
4. Rosales, R., Sclaroff, S.: Inferring body pose without tracking body parts. In: Proceedings of Computer Vision and Pattern Recognition, pp. 721–727 (2000)
5. Shakhnarovich, G., Viola, P., Darrell, T.: Fast pose estimation with parameter-sensitive hashing. In: Proceedings of International Conference on Computer Vision, pp. 750–757 (2003)
6. Gall, J., Rosenhahn, B., Brox, T., Seidel, H.P.: Optimization and filtering for human motion capture. International Journal of Computer Vision (2008)
7. Weik, S., Liedtke, C.E.: Hierarchical 3D pose estimation for articulated human body models from a sequence of volume data. In: Klette, R., Peleg, S., Sommer, G. (eds.) RobVis 2001. LNCS, vol. 1998, pp. 27–34. Springer, Heidelberg (2001)
8. Palágyi, K., Kuba, A.: A parallel 3D 12-subiteration thinning algorithm. Graphical Models and Image Processing 61(4), 199–221 (1999)

9. Pudney, C.: Distance-ordered homotopic thinning: A skeletonization algorithm for 3D digital images. In: Computer Vision and Image Understanding, vol. 72, pp. 404–413 (1998)
10. Arata, M., Kazuhiko, S., Takashi, M.: Human pose estimation from 3D object skeleton using articulated cylindrical human model. IPSJ SIG Technical Reports 51, 133–144 (2006)
11. Yang, H.D., Lee, S.W.: Reconstructing 3D human body pose from stereo image sequences using hierarchical human body model learning. In: ICPR 2006: Proceedings of the 18th International Conference on Pattern Recognition, Washington, DC, USA, pp. 1004–1007. IEEE Computer Society, Los Alamitos (2006)
12. Yang, H.D., Lee, S.W.: Reconstruction of 3D human body pose from stereo image sequences based on top-down learning. Pattern Recognition 40(11), 3120–3131 (2007)
13. Knoop, S., Vacek, S., Dillmann, R.: Fusion of 2D and 3D sensor data for articulated body tracking. Robotics and Autonomous Systems 57(3), 321–329 (2009)
14. Zhu, Y., Dariush, B., Fujimura, K.: Controlled human pose estimation from depth image streams. In: IEEE Computer Society Conference on Computer Vision and Pattern Recognition Workshops, 2008. CVPRW 2008, June 2008, pp. 1–8 (2008)
15. Böhme, M., Haker, M., Martinetz, T., Barth, E.: A facial feature tracker for human-computer interaction based on 3D TOF cameras. In: Dynamic 3D Imaging – Workshop in Conjunction with DAGM (2007) (in print)
16. Haker, M., Böhme, M., Martinetz, T., Barth, E.: Deictic gestures with a time-of-flight camera. In: Gesture in Embodied Communication and Human-Computer Interaction – International Gesture Workshop GW 2009 (2009)
17. Otsu, N.: A threshold selection method from gray-level histograms. IEEE Transactions on Systems, Man and Cybernetics 9(1), 62–66 (1979)
18. Martinetz, T., Schulten, K.: A "Neural-Gas" Network Learns Topologies. Artificial Neural Networks I, 397–402 (1991)

Analysis of Gait Using a Treadmill and a Time-of-Flight Camera

Rasmus R. Jensen, Rasmus R. Paulsen, and Rasmus Larsen

Informatics and Mathematical Modelling, Technical University of Denmark
Richard Petersens Plads, Building 321, DK-2800 Kgs. Lyngby, Denmark
{raje,rrp,rl}@imm.dtu.dk
www.imm.dtu.dk

Abstract. We present a system that analyzes human gait using a treadmill and a *Time-of-flight* camera. The camera provides spatial data with local intensity measures of the scene, and data are collected over several gait cycles. These data are then used to model and analyze the gait. For each frame the spatial data and the intensity image are used to fit an articulated model to the data using a Markov random field. To solve occlusion issues the model movement is smoothened providing the missing data for the occluded parts. The created model is then cut into cycles, which are matched and through Fourier fitting a cyclic model is created. The output data are: Speed, Cadence, Step length and *Range-of-motion*. The described output parameters are computed with no user interaction using a setup with no requirements to neither background nor subject clothing.

Keywords: Time-of-flight camera, Markov random fields, gait analysis, computer vision, motion capture.

1 Introduction

When computer vision is used in the study of biomechanics and motion capture it often involves complex setups. Different elements such as multiple cameras, bluescreens, markers and manual annotation along with a calibrated reference system are core in these setups. Obviously the more controlled the environment and the more sophisticated the setup the better the precision and thus, resulting models. This happens at an accordingly higher price and the complexity of the setups narrows down the application areas. Because of this, several approaches aim to simplify the tracking of movement.

Medina-Carnicer et al. [8] propose an algorithm that improves on current automatic detection of markers. This removes the tedious work of manually annotating markers in each frame.

Using several cameras but without bluescreens or markers Wan et al. [11] create a visual hull in space from silhouettes by solving a spatial Markov random field using graph cuts, then fit a model to this hull.

Based on a large database Shakhnarovich et al. [9] find a pose estimate in sublinear time relative to the database size. This algorithm uses subsets of features to find the nearest match in the parameter space.

R. Koch and A. Kolb (Eds.): Dyn3D 2009, LNCS 5742, pp. 154–166, 2009.

An earlier study by Zhu et al. [13] uses the *Time-of-flight (TOF)* camera estimate pose using key feature points in combination with an articulated model to solve problems with ambiguous feature detection, self penetration and joint constraints.

This article aims to deliver a system that analyzes gait without the use of the before mentioned core elements in the motion capturing.

We propose an adaptation of the *Posecut* algorithm by Bray et al. [5] for fitting articulated human models to greyscale image sequences to fitting such models to *TOF* camera sequences of spatial data with corresponding intensity. In particular, we will investigate the use of this *TOF* data adapted *Posecut* algorithm to quantitative gait analysis.

This article is on the progression of our previous work [6], where a simple model was used to analyze normal gait with a *TOF* camera. We have used a newer version of the *TOF* camera (the SwissRanger^{TM}SR4000 compared to the SR3000[2]), which improves on issues of reflection and precision. We also benefit from using a treadmill providing both a better relative subject to pixel resolution and allows for analysis of numerous steps in the gait instead of just a few when passing the camera. On a treadmill there is no actual movement in space, which changes the analysis.

The system requires no user interaction and there are no restrictions on neither background nor subject clothing and the motivation is to provide a system for physiotherapists simple enough that it could broaden the range of patients benefiting from an algorithmic gait analysis.

2   Time-of-Flight Data

The imaging device used is a SwissRanger^{TM}SR4000 [2], which emits a modulated signal from an array of near infrared LEDs surrounding the camera lens. By measuring the phase shift between the emitted and the recorded signal a depth map is calculated. The modulated signal has a wavelength of 10 m, which allows for distance measures of up to 5 m without ambiguity (10 m back and forth). The camera accuracy of the measured distance is less than 1 cm. Knowing the lens properties and a pixelwise depth the camera also creates a vertical and a horizontal map completing the spatial coordinates of the scene. To create an intensity image the camera uses the amplitude of the modulated signal. Because the signal from the LEDs and not the ambient light is used, the amplitude is relative to the square of the distance. An intensity map is provided being the amplitude times the squared distance as a correction. Figure 1 shows a depth map with intensity coloring; the vertical and horizontal maps are not shown.

3   Introduction to the Algorithm Finding the Pose

To find the pose of the subject in every frame in a sequence an adaption of the *Posecut* algorithm is used on the depth and intensity stream from *TOF* camera.

Fig. 1. Depth map with intensity coloring of the scene. The image is rotated to emphasize the spatial properties.

The algorithm uses 4 terms to define an energy minimization problem and find the pose of the subject as well as segmenting between subject and background:

Likelihood term: This term is based on statistics of the background using a probability function of a given pixel being labeled background.

Smoothness prior: Neighbouring pixels are expected to have the same label with higher probability than having different labels.

Contrast term: Neighbouring pixels with different labels are expected to have values in the intensity map that differs from one another. If the values are very similar but the labels different, this is penalized.

Shape prior: Trying to find the pose of a human, a human shape is used as a prior.

3.1 Random Fields

A frame in the sequence is considered to be a random field. A random field consists of a set of discrete random variables $\{X_1, X_2, \ldots, X_n\}$ defined on the index set I. In this set each variable X_i takes a value x_i from the label set $L = \{L_1, L_2, \ldots, L_k\}$ presenting all possible labels. All values of x_i, $\forall i \in I$ are represented by the vector $\mathbf{x}$ which is the configuration of the random field and takes values from the label set L^n. In the following the labeling is a binary problem, where $L = \{s, b\}$, where s is subject and b is background.

A neighbourhood system to X_i is defined as $N = \{N_i | i \in I\}$ for which it holds that $i \notin N_i$ and $i \in N_j \Leftrightarrow j \in N_i$. A random field is said to be a Markov field, if the probability P of any configuration of $\mathbf{x}$ satisfies the positivity property:

$$P(\mathbf{x}) > 0 \qquad \forall \mathbf{x} \in L^n \tag{1}$$

And the Markovian property:

$$P(x_i|\{x_j : j \in I\backslash\{i\}\}) = P(x_i|\{x_j : j \in N_i\}) \tag{2}$$

Or in other words the probability of x_i given the index set $I\backslash\{i\}$ is the same as the probability given the neighbourhood of i.

3.2 The Likelihood Function

The likelihood energy is based on the negative log likelihood and for the background distribution defined as:

$$\Phi(\mathbf{D}|x_i = \mathrm{b}) = -\log P(\mathbf{D}|x_i) \tag{3}$$

Where $\mathbf{D}$ is the observed data. Using the Gibbs measure on a background distribution with mean μ_b and standard deviation σ_b without the normalization constant the energy becomes:

$$\Phi(\mathbf{D}|x_i = \mathrm{b}) = \frac{(\mathbf{D} - \mu_{\mathrm{b},i})^2}{\sigma_{\mathrm{b},i}^2} \tag{4}$$

With no distribution defined for pixels belonging to the subject, the subject likelihood function is set to the mean of the background likelihood function. To estimate a stable background a variety of methods are available. A well known method models each pixel as a mixture of Gaussians and is also able to update these estimates on the fly [10]. When using a treadmill the subject stays in the same place and to avoid subject being modelled as background the background is simply estimated by computing the median and standard deviation of each pixel over a number of frames before the subject enters the scene.

3.3 The Smoothness Prior

This term states that generally neighbours have the same label with higher probability. The generalized Potts model where $j \in N_i$ is given by:

$$\psi(x_i, x_j) = \begin{cases} K_{ij} & x_i \neq x_j \\ 0 & x_i = x_j \end{cases} \tag{5}$$

This term penalizes neighbours having different labels. In the case of segmenting between background and subject, the problem is binary and referred to as the Ising model [4]. The parameter K_{ij} determines the smoothness in the resulting labeling.

3.4 The Contrast Term

In some areas such as where the feet touches the ground, the subject and background differs very little in distance. Therefore a contrast term is added, which uses the intensity image (grayscale) provided by the *TOF* camera. It is expected that two adjacent pixels with the same label have similar intensities, which implies that adjacent pixels with different labels have different intensities. By decreasing the cost of neighbouring pixels with different labels exponentially with an increase in difference in intensity, this term favours neighbouring pixels with similar intensities to have the same label. This function is defined as:

$$\gamma(i,j) = \lambda \exp\left(\frac{-g^2(i,j)}{2\sigma_{b,i}^2}\right) \tag{6}$$

Where $g^2(i,j)$ is the gradient in the intensity map and approximated using convolution with gradient filters. The parameter λ controls the cost of the contrast term, and the contribution to the energy minimization problem becomes:

$$\Phi(\mathbf{D}|x_i, x_j) = \begin{cases} \gamma(i,j) & x_i \neq x_j \\ 0 & x_i = x_j \end{cases} \tag{7}$$

3.5   The Shape Prior

To ensure that the segmentation is human like and wanting to estimate a human pose, a human shape model consisting of ellipses and circles is used as a prior. The model has 19 degrees of freedom coming from the position, the height and the joint angles. The model is 2-dimensional and to compensate for the difference of the legs in the frame due to difference in length to the focal point, the legs are allowed to stretch. The only restriction in the model is that the knee joints cannot overstretch. The hip joint is simplified such that the hip is connected in one point as studies shows that a 2-dimensional model can produce good results in gait analysis [3].

Pixels near the shape model in a frame are more likely to be labeled subject, while pixels far from the shape are more likely to be background.

The cost function for the shape prior is defined as:

$$\Phi(x_i|\Theta) = -\log(P(x_i|\Theta)) \tag{8}$$

Where Θ contains the pose parameters of the shape model (position, height, joint angles and stretch of legs). The probability $P(x_i|\Theta)$ of labeling subject or background is defined as follows:

$$P(x_i = \text{s}|\Theta) = 1 - P(x_i = \text{b}|\Theta) = \frac{1}{1 + \exp(\mu * (\text{dist}(i,\Theta) - d_r))} \tag{9}$$

The function $\text{dist}(i,\Theta)$ is the distance from pixel i to the shape defined by Θ, d_r is the width of the shape, and μ is the magnitude of the penalty given to points outside the shape. To calculate the distance for all pixels to the model,

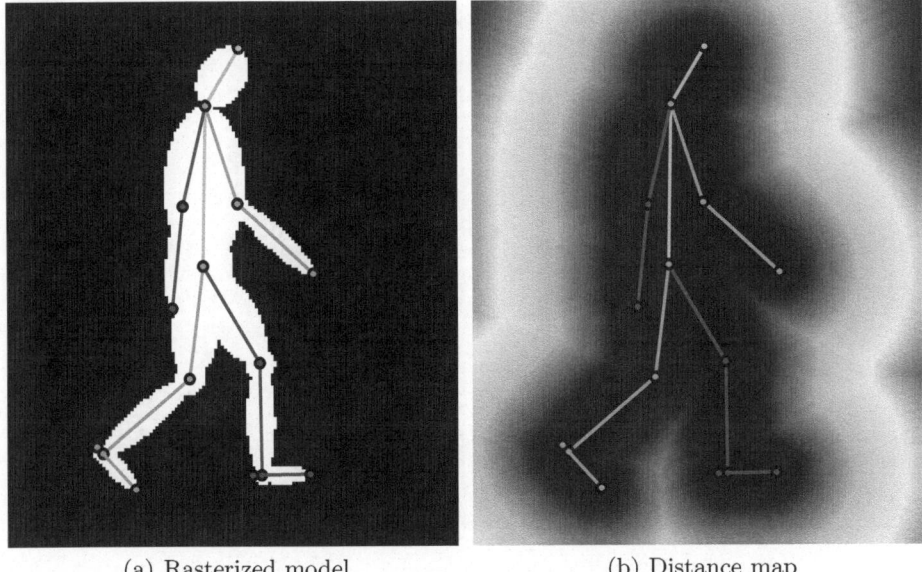

(a) Rasterized model   (b) Distance map

Fig. 2. Raster model and the corresponding distance map

the shape model is rasterized and the distance found using the *Signed Euclidian Distance Transform (SEDT)* [12]. Figure 2 shows the rasterized model with the skeletal structure and the corresponding distance map.

3.6   Energy Minimization

Combining the four energy terms a cost function for the pose and segmentation becomes:

$$\Psi(\mathbf{x}, \mathbf{\Theta}) = \sum_{i \in V} \left(\Phi(\mathbf{D}|x_i) + \Phi(x_i|\mathbf{\Theta}) + \sum_{j \in N_i} \left(\psi(x_i, x_j) + \Phi(\mathbf{D}|x_i, x_j) \right) \right) \quad (10)$$

This Markov random field is solved using *Graph Cuts* [7], and the pose is optimized in each frame using the pose from the previous frame as initialization.

3.7   Initialization

To find an initial frame and a pose, a frame that differs a lot from the background model is chosen and the difference is summed along the rows and columns. As a rough guess on where the subject is in this frame, these two sum vectors are used to guess the first and last rows and columns that contains the subject (Fig 3(a)). From the initial guess the pose is optimized according to the energy problem by searching locally. Figure 3(b) shows the optimized pose, where the limbs are

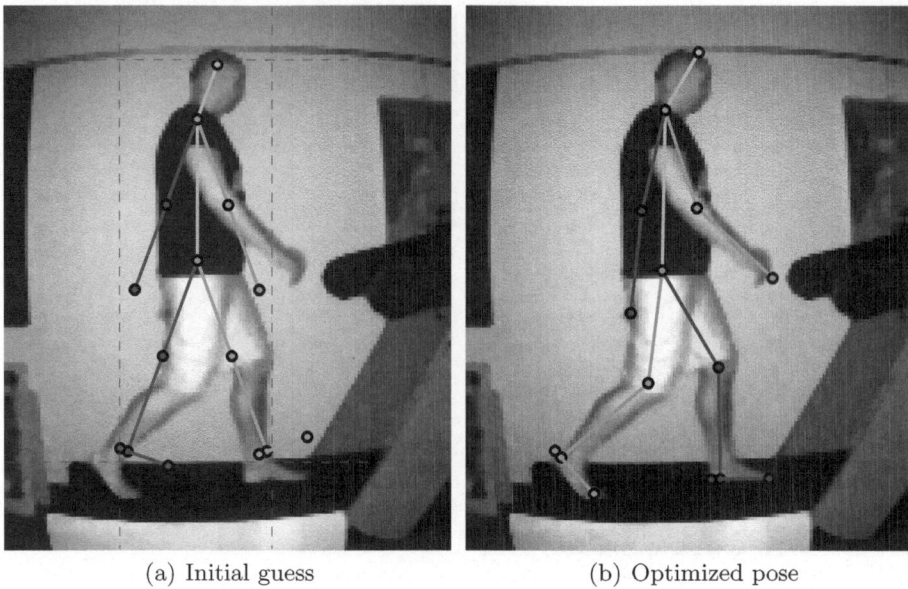

(a) Initial guess   (b) Optimized pose

Fig. 3. Initialization of the algorithm

differentiated based on distance measures to track left and right as described later.

The pose in the remaining frames is found using the previous frame as an initial guess and then optimizing on this. This generally works very well, but sometimes problems arise when the legs pass each other as feet or knees of one leg tend to get stuck on the wrong side of the other leg. This entanglement is avoided by not allowing crossed legs as an initial guess and instead using straight legs close together. When both arms end up in front or behind the torso this constitutes a local minimum in the energy minimization. This is avoided by letting one arm lead and one arm follow in the initialization of each frame.

4   Postprocessing

Because the algorithm is initialized using a rough guess, there is no initial knowledge of left and right in the model. The Markov random field used to find the pose is solved using a binary solution, so this does not provide information about left and right. Also a few problems occur when limbs are occluded by other limbs. Therefore some postprocessing must be done to keep track of left and right and to ensure a smooth movement.

4.1   Tracking Left and Right

When the subject walks facing right in the frame, the right side of the subject will always be closer to the camera in the depth map. By ensuring that this

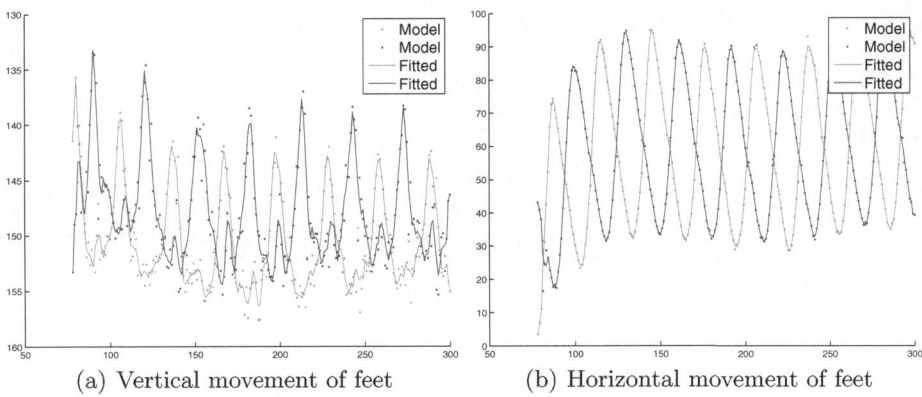

(a) Vertical movement of feet   (b) Horizontal movement of feet

Fig. 4. 4(a) shows the vertical movement of the foot joints using image notation, where rows are increased downwards. 4(b) shows the horizontal movement. The scales of the two plots are not the same, and the vertical movement is much smaller than the horizontal.

holds true for the median of depth of the rasterized model limbs the tracking of left and right is constituted.

4.2   Smoothing the Movement

With the tracking of left and right the subject pose is now estimated in every frame. Our previous study [6] shows that using a smoothness prior on the movement, can improve on the results. The prior that movement is locally smooth holds true, if the framerate is high enough relative to the movement. A mean filtering and a piecewise polynomial fitting is used to smoothen the movement. Figure 4 shows the foot joint movement and the fitting.

5   Analyzing the Gait

With a smooth subject movement, the model is now ready for analysis.

5.1   Output Parameters

With the pose estimated in every frame the gait can now be analyzed. To find the steps during gait, the frames where the distance between the feet has a local maximum are used. Combining this with information about which foot is leading, the foot that is taking a step can be found. From the provided Cartesian coordinates in space the step lengths are found (Fig. 5(b) and 5(d)). By aligning the shape throughout the sequence the averages and standard deviations of the

(a) Left start   (b) Left stop   (c) Right start   (d) Right stop

Fig. 5. Step lengths: Left 0.60 m and Right 0.64 m

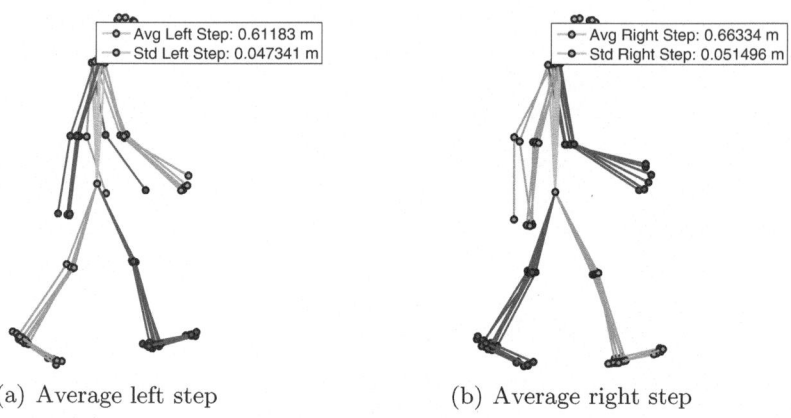

(a) Average left step   (b) Average right step

Fig. 6. Average step lengths

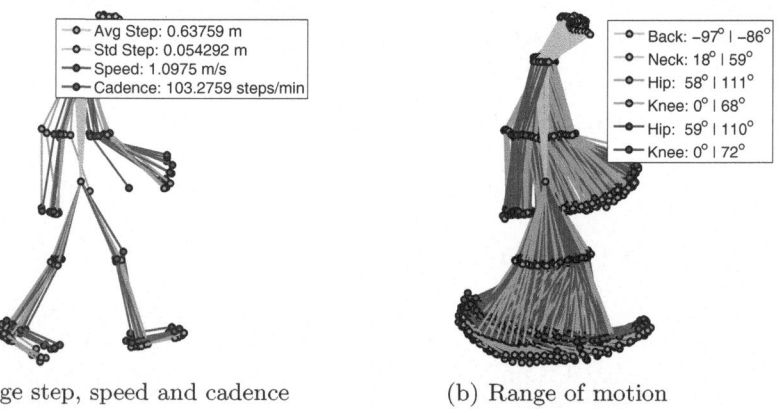

(a) Average step, speed and cadence   (b) Range of motion

Fig. 7. Several gait parameters and the skeletal cyclic alignment

steps are also found as shown in Fig 6. Notice that the step lengths are almost identical as expected with a normal persons gait. With a timestamp for each frame the speed and cadence are found (Fig. 7(a)). The *range-of-motion* is found using the minima and maxima of the joints (Fig. 7(b)).

The walking speed found by the algorithm matches the actual speed of the treadmill very well. The speed found in Fig. 7(a) is 1.10 m/s equal to 3.96 km/h, and the treadmill was set to 4 km/h. This is a good measure of the correctness of the system, as the speed is calculated as a sum of all steps in a sequence over time, and therefore not expected to fluctuate a lot. Measuring the speed in several sequences showed a very high correlation between the estimated speed and that of the treadmill.

The range of motion is found as the clockwise angle from the x-axis in positive direction for the inner limbs (femurs and torso) and the clockwise change compared to the inner limbs for the outer joints (ankles and head). Figure 7(b) shows the angles and the model pose throughout the sequence, as expected with person with normal gait the range of motion is almost identical for both legs.

5.2   A Cyclic Model

The movement is cut into cycles and the shapes aligned such that the different phases of the cycles match. Figure 8 shows 4 phases of the cyclic alignment used

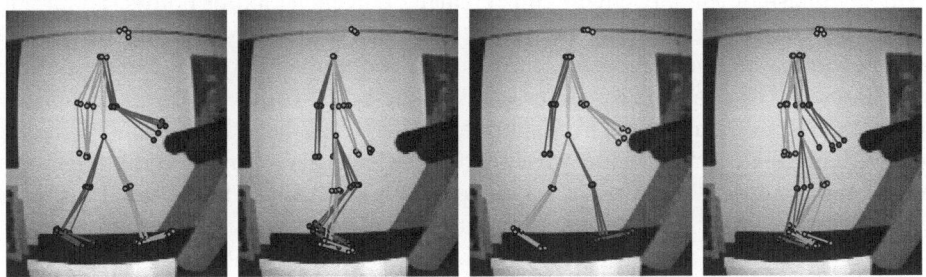

Fig. 8. Phase aligned shapes with the hip as the center of mass.

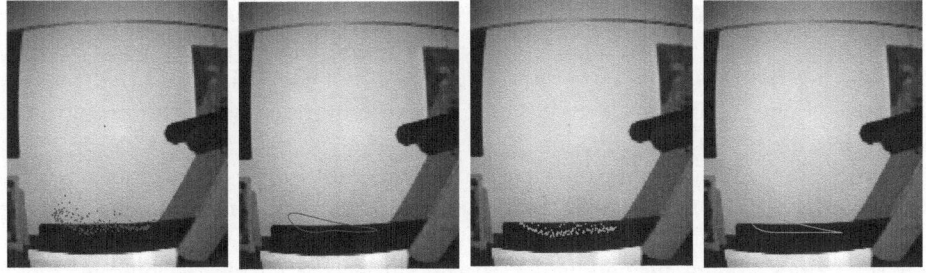

Fig. 9. Aligned points and cyclic movement of the feet

Fig. 10. Model of the subjects gait cycle

in the modeling. Every point in the model is aligned relative to both the hip and the phase of the movement thus creating a cyclic cloud of points. By fitting Fourier series to these points the point clouds can be modeled as cycles (Fig. 9 shows the aligned movement of the feet and the modeling of the feet). Looking at the power of the Fourier series shows that a 9th order series makes a very good model, and that very little happens at frequencies described by a higher order. The alignment is done with the hip as a center of mass and therefore a cyclic movement of the hip has to be added to finalize the modeling. Figure 10 shows the the cyclic model in every phase plotted in the same figure.

6   Conclusion

A system is created that analyzes gait. Using a treadmill the system does the analysis over several gait cycles producing several gait parameters as well as a cyclic model of the subjects gait. The camera used is a SwissRanger^{TM}SR4000 [2], which provides spatial data as well as intensity images. Because the tracking is mainly done in the depth map, there are no requirements to the background or to the subject clothing.

The system also initializes itself removing the need for user interaction.

Even though the model used is 2-dimensional the output is 3-dimensional, the joint angles however are solely computed in the plane and therefore dependent on the camera being perpendicular to the subject. It seems reasonable that a single camera setup would always benefit from a camera angle perpendicular to the gait direction.

Visual inspection of the algorithm output shows a good correlation between model joints and the subject joints in the images. Our previous studies [6] has shown that it performs well compared to human annotation. The estimated subject speed matches that of the treadmill very well. For further testing the system output should be held against that of a full commercial setup.

The system has been implemented in MATLAB using C++ mex–files for the central parts, it computes a frame in a few seconds making realistically useable but not exactly realtime.

Acknowledgements

This work was in part financed by the ARTTS [1] project (Action Recognition and Tracking based on Time-of-Flight Sensors) which is funded by the European Commission (contract no. IST-34107) within the Information Society Technologies (IST) priority of the 6th framework Programme. This publication reflects only the views of the authors, and the Commission cannot be held responsible for any use of the information contained herein.

References

[1] Artts (2009), http://www.artts.eu
[2] Mesa (2009), http://www.mesa-imaging.ch
[3] Alkjaer, E.B., Simonsen, T., Dygre-Poulsen, P.: Comparison of inverse dynamics calculated by two- and three-dimensional models during walking. In: 2001 Gait and Posture, pp. 73–77 (2001)
[4] Besag, J.: On the statistical analysis of dirty pictures. Journal of the Royal Statistical Society. Series B (Methodological) 48(3), 259–302 (1986)
[5] Bray, M., Kohli, P., Torr, P.H.S.: Posecut: simultaneous segmentation and 3D pose estimation of humans using dynamic graph-cuts. In: Leonardis, A., Bischof, H., Pinz, A. (eds.) ECCV 2006. LNCS, vol. 3952, pp. 642–655. Springer, Heidelberg (2006)
[6] Jensen, R.R., Paulsen, R.R., Larsen, R.: Analyzing gait using a time-of-flight camera. In: Salberg, A.B., Hardeberg, J.Y., Jenssen, Q. (eds.) SCIA 2009. LNCS, vol. 5575, pp. 21–30. Springer, Heidelberg (2009)
[7] Kolmogorov, V., Zabin, R.: What energy functions can be minimized via graph cuts? IEEE Transactions on Pattern Analysis and Machine Intelligence 26(2), 147–159 (2004)
[8] Medina-Carnicer, R., Garrido-Castro, J.L., Collantes-Estevez, E., Martinez-Galisteo, A.: Fast detection of marker pixels in video-based motion capture systems. Pattern Recognition Letters 30(4), 432–439 (2009)
[9] Shakhnarovich, G., Viola, P., Darrell, T.: Fast pose estimation with parameter-sensitive hashing. In: Proceedings Ninth IEEE International Conference on Computer Vision, vol. 2, pp. 750–757 (2003)

[10] Stauffer, C., Grimson, W.E.L.: Adaptive background mixture models for real-time tracking. In: Proceedings. 1999 IEEE Computer Society Conference on Computer Vision and Pattern Recognition (Cat. No PR00149), vol. 2, pp. 246–252 (1999), ISSN 10636919

[11] Wan, C., Yuan, B., Miao, Z.: Markerless human body motion capture using Markov random field and dynamic graph cuts. Visual Computer 24(5), 373–380 (2008)

[12] Ye, Q.-Z.: The signed Euclidean distance transform and its applications. In: 1988 Proceedings of 9th International Conference on Pattern Recognition, vol. 1, pp. 495–499 (1988)

[13] Zhu, Y., Dariush, B., Fujimura, K.: Controlled human pose estimation from depth image streams. In: 2008 IEEE Computer Society Conference on Computer Vision and Pattern Recognition Workshops (CVPR Workshops), pp. 1–8 (2008)

Face Detection Using a Time-of-Flight Camera

Martin Böhme, Martin Haker, Kolja Riemer,
Thomas Martinetz, and Erhardt Barth

Institute for Neuro- and Bioinformatics, University of Lübeck
Ratzeburger Allee 160, D-23538 Lübeck, Germany
{boehme,haker,riemer,martinetz,barth}@inb.uni-luebeck.de
http://www.inb.uni-luebeck.de

Abstract. We adapt the well-known face detection algorithm of Viola
and Jones [1] to work on the range and intensity data from a time-of-
flight camera. The detector trained on the combined data has a higher
detection rate (95.3%) than detectors trained on either type of data alone
(intensity: 93.8%, range: 91.2%). Additionally, the combined detector
uses fewer image features and hence has a shorter running time (5.15 ms
per frame) than the detectors trained on intensity or range individually
(intensity: 10.69 ms, range: 5.51 ms).

1 Introduction

In this paper, we will examine how a time-of-flight camera can be used for face
detection. We will extend the well-known face detection algorithm of Viola and
Jones [1] to time-of-flight (TOF) images; as we will show in the results section,
the detector trained on the combined range and intensity data not only has a
higher detection rate than detectors trained on either type of data alone, but it
also requires fewer features and therefore has a shorter running time.

The Viola-Jones face detector is computationally very efficient while at the
same time achieving good detection rates. This is due to three important characteristics: (i) The detector is based on image features that can be evaluated
quickly and in constant time, independent of the size of the feature; (ii) the
detector selects a set of highly discriminative image features using the *AdaBoost*
algorithm; (iii) the detector is structured into a cascade of progressively more
sophisticated stages. Since most candidate regions in an image are very dissimi-
lar to a face, the early stages of the cascade can discard these regions with little
computation; the later stages of the cascade, which require more computation,
need to process only a small proportion of candidate regions.

The attractive properties of the Viola-Jones face detector have motivated a
large number of researchers to extend this work in various ways, including the use
of different features, modifications to the AdaBoost algorithm, and the application to different types of object detection tasks (see e.g. [2,3,4]). The algorithm
has also already been applied to TOF data [5]. However, this previous work
does not extend the Viola-Jones detector itself to use range features; instead,
a standard Viola-Jones detector trained on images from a conventional camera

R. Koch and A. Kolb (Eds.): Dyn3D 2009, LNCS 5742, pp. 167–176, 2009.
© Springer-Verlag Berlin Heidelberg 2009

is used to find candidate faces in the TOF image; a final detector stage then computes the average distance of each candidate from the range map and rejects candidates whose size does not match the expected size of a face at this distance.

In contrast, the approach we will use in this paper is to include range features as well as intensity features in the set of features used by the detector. As we will show in the results section, the features chosen by the resulting detector consist of an approximately equal number of range and intensity features; the detector has a higher detection rate and shorter running time than detectors trained on the same training samples, but using either the range or the intensity information alone. This underlines previous results (see e.g. [6,7]) showing that it is the combination of range and intensity data that makes the TOF camera a valuable tool for object detection tasks.

2   Method

We use the basic face detection method of Viola and Jones [1] (which we will summarize briefly) but extend the set of features used to both range and intensity features. Since the method was first described, a number of authors have made improvements to the method (see e.g. [2,3,4]), but we use the original algorithm here because we are more interested in the difference made by using range data rather than in absolute performance.

The Viola-Jones face detector consists of a cascade of stages that typically become more sophisticated as one progresses through the cascade (see Fig. 1). The idea is that the overwhelming majority of subregions in an image are non-faces, and that most of these subregions are "easy", i.e. they can be identified as nonfaces with little computation. Thus, the first stage of the detector contains a computationally efficient classifier that can immediately reject most subregions as being nonfaces; no further processing is carried out on these subregions. Only a small fraction of subregions (both true faces and "hard" nonfaces) are passed on to the next stage for further processing. This next stage performs more computation and, by doing so, can again reject most of the subregions as being nonfaces, passing only a small fraction of subregions on to the next stage, and so on. In this way, the average effort per subregion is kept low because the overwhelming majority of subregions are rejected in the first few stages.

If the detection rate and false-positive rate of the i-th stage (on the input it receives from the previous stage) are d_i and f_i, then the overall detection and false-positives rates of an n-stage cascade are $D = \prod_{i=1}^{n} d_i$ and $F = \prod_{i=1}^{n} f_i$, respectively. A common approach is to train each stage to achieve the same detection rate d and false-positive rate f on its respective input; this results in overall detection and false-positive rates of $D = d^n$ and $F = f^n$.

Each cascade stage is a boosted classifier trained using the AdaBoost algorithm [8]; a boosted classifier combines several *weak classifiers* (each of which performs only slightly better than chance) into a *strong classifier* (which performs substantially better than the individual weak classifiers). The weak

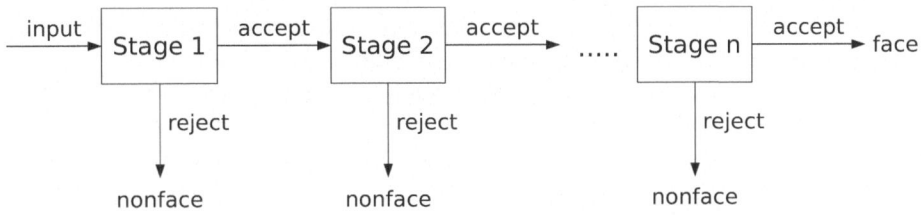

Fig. 1. Cascade structure of the Viola-Jones face detector

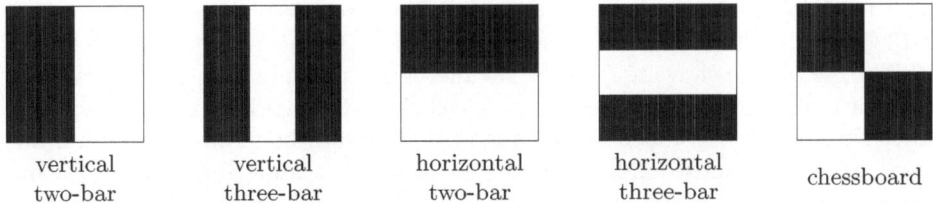

| vertical two-bar | vertical three-bar | horizontal two-bar | horizontal three-bar | chessboard |

Fig. 2. Haar-like features used by the Viola-Jones face detector. The feature value is obtained by summing the pixels in the white rectangle(s), then subtracting the sum of pixels in the black rectangle(s).

classifiers in the Viola-Jones algorithm are obtained by applying a threshold to an image feature.

The image features, finally, are composed of adjacent rectangles (see Fig. 2); the pixels within each rectangle are summed together, and the resulting values are added or subtracted to obtain the final feature value. For example, the value of the "vertical two-bar" feature is obtained by summing the pixels in the white rectangle and subtracting the sum of pixels in the black rectangle. These features (which are often called *Haar-like features*) have the advantage that they can be evaluated in constant time, independent of their size, using a data structure known as an *integral image*.

The feature set for training the detector is obtained by scaling these features to all possible widths and heights and translating them to all possible positions in the image. Also, each feature (at each size and position) may be evaluated either on the range data or on the intensity data.

Training of the cascade now proceeds as follows. We begin with a training set of face and nonface image patches of constant size. These are used to train the first cascade stage to the desired detection and false-positive rate (evaluated on a validation set). Now, because the next stage will never see those nonface patches that the first stage rejects, we discard all nonface samples rejected by the first stage from the training and validation set, keeping only the false positives. To bring the training and validation set back to their original sizes, we generate new nonface samples by scanning the cascade that has been trained so far across a set of images not containing faces and adding those subregions that the cascade erroneously classifies as faces to the training or validation set until both have been replenished. We continue adding stages to the cascade in this way until the false-positive rate of the cascade reaches a set target.

Detection proceeds by scanning the cascade across the input image in steps of a certain size. To be able to detect faces of different sizes, the subwindow processed by the detector, along with the features contained in it, is progressively scaled up by a certain factor until it reaches the size of the complete image.

3   Results

The training data for the face detector were recorded using a SwissRanger SR3000 camera [9]. The training set consists of 1310 images (with a resolution of 176 by 144 pixels) showing faces of 17 different persons, in different orientations and with different facial expressions, as well as 4980 images not containing faces. Each face image was labelled by hand with a square bounding box containing the face; some background was included in the bounding box, since previous researchers had reported that this yielded slightly better results than a more tighly cropped bounding box (see the discussion in [1, Sect. 5.1]).

The images were split up into a training, validation and test set, containing 70%, 23%, and 7% of the images, respectively. (The training set is used to select the best weak classifiers for each cascade stage, the validation set is used to evaluate whether the stage has reached its goal detection rate and false-positive rate, and the test set is used to test the final cascade after training is completed.) Face images were cropped to the face bounding box and resized to 24 by 24 pixels (see Fig. 3 for examples). To increase the number of face images in each set, we added versions of each image that were rotated left and right by 3 degrees. After this step, a mirrored version of each face image (including the rotated ones) was also added to the set. The nonface images were full frames of 176 by 144 pixels; to generate examples for training the first cascade stage, subimages of 24 by 24 pixels were cut out of the nonface images. For the second and subsequent stages,

Fig. 3. Examples of intensity images from the face training set

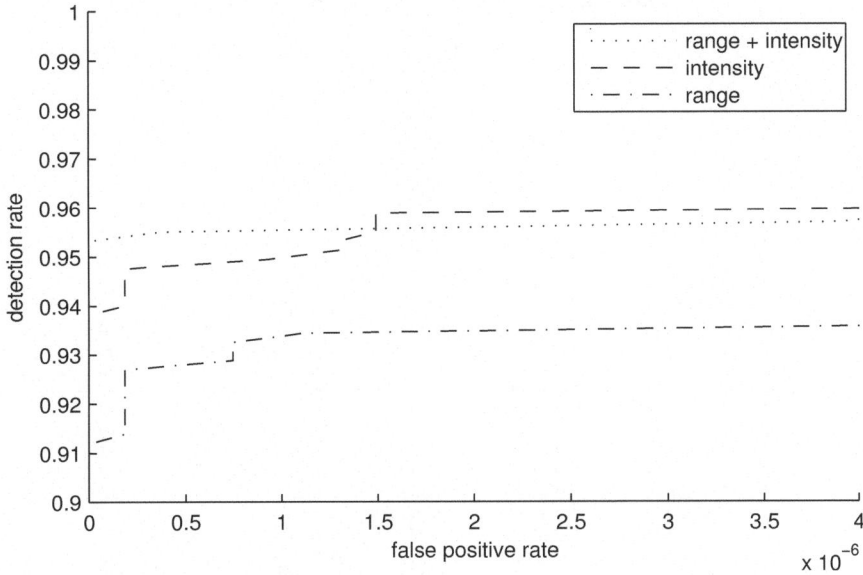

Fig. 4. ROC curves for the detectors trained on the combined range and intensity data as well as both types of data separately

new negative examples were generated by scanning the cascade trained so far across the nonface images and collecting false positives (see also Sect. 2).

In all, there were 5412 faces and 3486 nonface images in the training set, 1752 faces and 1145 nonface images in the validation set, and 534 faces and 349 nonface images in the test set. The data set is publicly available at www.artts.eu/publications/3d_tof_db

We trained a detector on the combined range and intensity data as well as on the range and intensity data alone. The target detection rate and false-positive rate for each stage were set to $d = 0.995$ and $f = 0.4$, respectively; the target false positive rate for the complete detector was set to 10^{-8}. The range-and-intensity detector as well as the range-only detector were successfully trained to this target rate. Training of the intensity-only detector did not reach the target rate; training was stopped manually when the detector had added over 1500 features to the cascade stage it was training and the false-positive rate of the stage had stagnated without reaching the goal rate. (This is a typical sign that the detector can no longer generalize from the training to the validation set.) We trained another intensity detector with a lower detection rate per stage of $d = 0.99$; training for this detector did complete, but its performance was consistently worse than that of the detector with $d = 0.995$ whose training was aborted. For this reason, we will only use the latter detector in the tests that follow.

Figure 4 shows ROC curves for the three detectors (computed as in [3]). For false positive rates above $1.5 \cdot 10^{-6}$, the intensity-only detector achieves a slightly higher detection rate than the range-and-intensity detector. Below

Table 1. Cascade structure of the detectors

| | Stage | detection rate | | false-positive rate | | number of features | | |
		individual	cumulative	individual	cumulative	total	intensity	range
intensity + range	0	1.000	1.000	0.000	1.0e-02	2	1	1
	1	0.995	0.995	0.165	1.6e-03	2	1	1
	2	0.995	0.991	0.138	2.3e-04	2	1	1
	3	0.995	0.986	0.394	8.9e-05	7	4	3
	4	0.995	0.981	0.290	2.6e-05	7	3	4
	5	0.995	0.976	0.385	1.0e-05	12	5	7
	6	0.995	0.971	0.334	3.3e-06	12	8	4
	7	0.995	0.966	0.381	1.3e-06	15	9	6
	8	0.995	0.961	0.380	4.8e-07	20	11	9
	9	0.995	0.956	0.380	1.8e-07	22	13	9
	10	0.995	0.951	0.368	6.8e-08	27	16	11
	11	0.995	0.946	0.365	2.5e-08	44	26	18
	12	0.995	0.941	0.391	9.6e-09	49	26	23
intensity	0	0.999	0.999	0.189	1.9e-01	3	3	0
	1	0.995	0.994	0.367	6.9e-02	15	15	0
	2	0.995	0.989	0.340	2.4e-02	11	11	0
	3	0.998	0.986	0.340	8.0e-03	6	6	0
	4	0.997	0.983	0.391	3.1e-03	10	10	0
	5	0.995	0.978	0.389	1.2e-03	17	17	0
	6	0.995	0.973	0.376	4.6e-04	25	25	0
	7	0.995	0.968	0.384	1.8e-04	23	23	0
	8	0.995	0.963	0.386	6.8e-05	40	40	0
	9	0.995	0.958	0.389	2.6e-05	59	59	0
	10	0.995	0.953	0.392	1.0e-05	62	62	0
	11	0.995	0.948	0.400	4.1e-06	107	107	0
	12	0.995	0.943	0.396	1.6e-06	198	198	0
	13	0.995	0.938	0.399	6.5e-07	117	117	0
	14	0.995	0.934	0.390	2.5e-07	223	223	0
range	0	0.997	0.997	0.071	7.1e-02	2	0	2
	1	0.995	0.992	0.285	2.0e-02	3	0	3
	2	1.000	0.992	0.325	6.5e-03	3	0	3
	3	0.995	0.987	0.373	2.4e-03	11	0	11
	4	0.995	0.983	0.271	6.6e-04	17	0	17
	5	0.995	0.978	0.374	2.5e-04	18	0	18
	6	0.996	0.974	0.381	9.4e-05	10	0	10
	7	0.995	0.969	0.388	3.6e-05	26	0	26
	8	0.995	0.964	0.377	1.4e-05	29	0	29
	9	0.995	0.959	0.396	5.4e-06	56	0	56
	10	0.995	0.954	0.370	2.0e-06	42	0	42
	11	0.995	0.949	0.381	7.7e-07	66	0	66
	12	0.995	0.944	0.396	3.0e-07	114	0	114
	13	0.995	0.940	0.382	1.2e-07	81	0	81
	14	0.995	0.935	0.377	4.4e-08	111	0	111
	15	0.995	0.930	0.380	1.7e-08	139	0	139

this point, the range-and-intensity detector achieves better detection rates. Both detectors are markedly better than the range-only detector over the whole range of false-positive rates shown. All three detectors achieve good detection rates even for a false-positive rate of zero. This is an indication that our test set is relatively "easy" compared to, for instance, the MIT+CMU test set [10], on which the Viola-Jones algorithm produces slightly higher error rates [1]. Whereas the MIT+CMU test set contains images from a variety of sources, including text and line drawings, our test set consists solely of images taken with a single camera. Also, because of the active illumination, the lighting is the same across all images. We believe these factors combine to make the test set "easier".

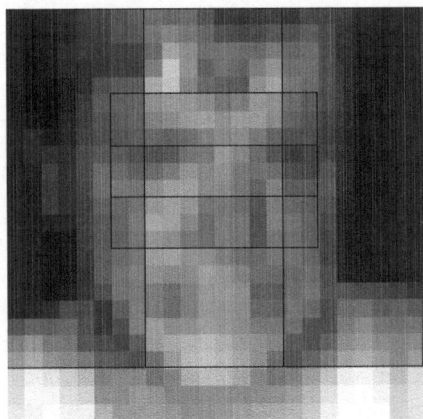

Fig. 5. Features used by the first stage of the range-and-intensity detector. The blue (vertical) feature is a range feature; the red (horizontal) feature is an intensity feature.

We will now examine the cascade structure of the detectors, i.e. the number of features used in each stage together with the detection rate and false-positive rate for each stage (see Table 1).

The first thing that is noticeable is that the first stage of the range-and-intensity detector achieves a false-positive rate of 0 on the validation set, i.e. the false-positive rate was too small to measure on the validation set. When this first stage was run on the set of full-frame test images, its false positive rate was 0.05%. In other words, the first stage already eliminates 99.95% of nonfaces. For comparison, the first stages of the other two detectors had false-positive rates of 18.9% (intensity) and 7.1% (range).

To understand why the first stage of the range-and-intensity detector has such good performance, consider Fig. 5, which shows the features used by this stage: A vertical three-bar range feature and a horizontal three-bar intensity feature. From the sample training image underlayed under the features, it is evident that the range feature responds to the range difference between the face and the background on either side; the intensity feature seems to respond to the difference between the eye region (which is typically darker) and the forehead and cheeks above and below (which are typically lighter).

The fact that the first stage achieves a false-positive rate of zero on the validation set is problematic for computing the false-positive rate of the entire cascade, which is used during training to decide when the detector has reached its performance goal. To be able to compute an overall false-positive rate, we conservatively assumed the false-positive rate for this stage to be 0.01; this assumption is also used in the cumulative rates shown in the table. The assumed rate of 0.01 is probably quite conservative and only affects the overall false-positive rate computed during training, but not the selection of weak classifiers or the false-positive rates computed on the test set.

Table 2. Performance summary of the detectors on the various types of data. Detection rates are given for a zero false-positive rate on the test set. Running times include preprocessing (computation of the integral images).

Detector	Detection rate	Running time per frame
range + intensity	95.3%	5.15 ms
intensity	93.8%	10.69 ms
range	91.2%	5.51 ms

Turning to the number of features per stage, we note that, in most stages, the range-and-intensity detector requires noticeably fewer features to reach its target performance than the other two detectors. Also, note that the range-and-intensity detector uses an approximately equal number of range and intensity features in each stage (with a tendency to use slightly more intensity features in the later stages). This indicates that the range and intensity data contribute approximately the same amount of information to the face detection task.

Finally, we turn to the running times for the various detectors (Table 2). The range-and-intensity detector is more than two times faster than the intensity-only detector and slightly faster than the range-only detector. This reflects the fact that the intensity-only detector uses more features than the other two detectors in the first few cascade stages, which consume the most processing time. Note when comparing the timings that the range-and-intensity detector needs to perform twice the amount of preprocessing (computing the integral images for both range and intensity) but still ends up faster. The table also summarizes the detection rates achieved for a zero false-positive rate on the test set.

4   Discussion

We have shown that a face detector trained on the combined range and intensity data from a TOF camera yields a higher detection rate (95.3%) than a detector trained on either type of data alone (intensity: 93.8%, range: 91.2%). Furthermore, the range-and-intensity detector requires fewer features than the other two detectors. This translates into faster running times: The range-and-intensity detector is over twice as fast as the intensity-only detector and slightly faster than the range-only detector (which misclassifies almost twice as many faces).

The data obtained by the TOF camera is in effect a two-channel image, where one channel contains the range map and the other contains the intensity image. If the TOF camera is combined with a grayscale or RGB camera operating in the visible spectrum (as is the case in the 3DV Systems ZCam [11], for instance), it would be straightforward to extend the method to the additional channels obtained in this way.

The detector we used was a "stock" Viola-Jones face detector. Even better results might be possible using features that are specifically tuned to the type of structures typically found in range images. One could also investigate the idea of combining range and intensity information in a single feature. Additionally, the many refinements that have been made to the Viola-Jones algorithm since its inception could be incorporated.

However, we are not primarily interested in the maximum absolute performance that a TOF face detector can achieve but rather in the relative difference in performance between face detection on combined range and intensity data versus either type of data alone. We believe that the advantage of the combined range and intensity detector in terms of robustness and speed should be preserved when refinements are made to the underlying algorithms; whether this indeed holds true is a question for future research.

Acknowledgments

We thank the anonymous reviewers for their comments, which helped to improve this paper. This work was developed within the ARTTS project (www.artts.eu), which is funded by the European Commission (contract no. IST-34107) within the Information Society Technologies (IST) priority of the 6th Framework Programme. This publication reflects the views only of the authors, and the Commission cannot be held responsible for any use which may be made of the information contained therein.

References

1. Viola, P., Jones, M.: Robust real-time face detection. International Journal of Computer Vision 57(2), 137–154 (2004)
2. Lienhart, R., Kuranov, A., Pisarevsky, V.: Empirical analysis of detection cascades of boosted classifiers for rapid object detection. In: Michaelis, B., Krell, G. (eds.) DAGM 2003. LNCS, vol. 2781, pp. 297–304. Springer, Heidelberg (2003)
3. Brubaker, S.C., Wu, J., Sun, J., Mullin, M.D., Rehg, J.M.: On the design of cascades of boosted ensembles for face detection. International Journal of Computer Vision 77(1–3), 65–86 (2008)
4. Barczak, A.L.C., Johnson, M.J., Messom, C.H.: Real-time computation of Haar-like features at generic angles for detection algorithms. Research Letters in the Information and Mathematical Sciences 9, 98–111 (2006)
5. Hansen, D.W., Larsen, R., Lauze, F.: Improving face detection with TOF cameras. In: Proceedings of the IEEE International Symposium on Signals, Circuits & Systems (ISSCS), vol. 1, pp. 225–228 (2007)
6. Haker, M., Böhme, M., Martinetz, T., Barth, E.: Geometric invariants for facial feature tracking with 3D TOF cameras. In: Proceedings of the IEEE International Symposium on Signals, Circuits & Systems (ISSCS), Iasi, Romania, vol. 1, pp. 109–112 (2007)
7. Haker, M., Böhme, M., Martinetz, T., Barth, E.: Scale-invariant range features for time-of-flight camera applications. In: CVPR 2008 Workshop on Time-of-Flight-based Computer Vision, TOF-CV (2008)

8. Freund, Y., Schapire, R.E.: A decision-theoretic generalization of on-line learning and an application to boosting. Journal of Computer and System Sciences 55(1), 119–139 (1997)
9. Oggier, T., Büttgen, B., Lustenberger, F., Becker, G., Rüegg, B., Hodac, A.: SwissRanger^{TM} SR3000 and first experiences based on miniaturized 3D-TOF cameras. In: Proceedings of the 1st Range Imaging Research Day, Zürich, Switzerland, pp. 97–108 (2005)
10. Rowley, H.A., Baluja, S., Kanade, T.: Neural-network-based face detection. IEEE Transactions on Pattern Analysis and Machine Intelligence 20(1), 23–38 (1998)
11. ZCam: 3DV Systems, Yokne'am, Israel, http://www.3dvsystems.com.

Author Index